CHAINSAWS IN TROPICAL FORESTS

CHAINSAWS
in tropical forests

a manual prepared by the

Food and Agriculture Organization of the United Nations

and the

International Labour Organisation

FOOD AND AGRICULTURE ORGANIZATION OF THE UNITED NATIONS
ROME 1980

The publication of this manual was made possible by a special contribution from Norway, under the FAO/NORWAY Cooperative Programme: TF/INT 283 (NOR).

P-39

ISBN 92-5-100932-5

TABLE OF CONTENTS

1. INTRODUCTION

FAO/ILO[1/] attach great importance to their educational and training activities as a means of transferring knowledge and technology to the developing countries. This training manual has been especially prepared for foresters, loggers, foremen and workers in developing countries. Its aim is to make the use of chainsaws easier, safer and more efficient in the felling, delimbing and cross-cutting of trees.

An attempt has been made to restrict this manual to basic techniques and basic knowledge, to use simple language and to explain as much as possible by drawings. The manual is issued in English, French and Spanish. Requests for authorization to publish it in other languages should be addressed to FAO, Rome. The illustrations may also be used for projection.

This manual does not replace the instruction manual provided by chainsaw manufacturers, which should, in addition, always be studied carefully.

It is not the object of this manual to promote the change-over from hand tools to chainsaws. There are many cases where the use of hand saws is preferable for social and economic reasons. Special training material also exists for hand tools[2/]. But at the same time chainsaws are increasingly being used and therefore also require adequate attention.

This manual has been made possible by a special contribution from Norway, under the FAO/Norway Cooperative Programme. It has been jointly prepared by FAO and ILO. It is largely based on "The Chainsaw - Use and Maintenance", published by the National Board of Forestry of Sweden; the ILO training manual "Felling and Cross-cutting of Tropical Trees in Natural Forests", and on the ILO "Code of Practice on Safe Design and Use of Chainsaws". Drafts for this manual were prepared by Arne Been, Philippe Crubile and Bernt Strehlke in collaboration with the Forest Logging and Transport Branch of FAO. Advice was given by logging specialists Romesh Chandra, Erling Fosser and Klaus Virtanen and the manual was illustrated by Nils Forshed. The project leader was Gunnar Segerström of FAO.

Any comments and suggestions with regard to modifications and improvements of this manual will be welcome.

1/ FAO: Food and Agriculture Organization of the United Nations,
Viale delle Terme di Caracalla, Roma 00100, Italy.

ILO: International Labour Organization,
4 route des Morillons, CH-1211 Geneva 22, Switzerland

2/ "Selection and Maintenance of Logging Hand Tools",
ILO, Geneva, 1970.

2. THE CHAINSAW

2.1 MAIN PARTS

Most chainsaws are similar in design. Some parts are the same for different models, for instance guide bars and chains.

The drawing on the opposite page shows the main parts of a chainsaw, which are:

1. Front handle
2. Saw chain
3. Guide bar
4. Spikes (optional in small-sized trees)
5. Choke
6. Rear handle
7. Throttle control latch
8. Throttle control trigger
9. On/Off control
10. Starter

Positions may vary from one machine to another.

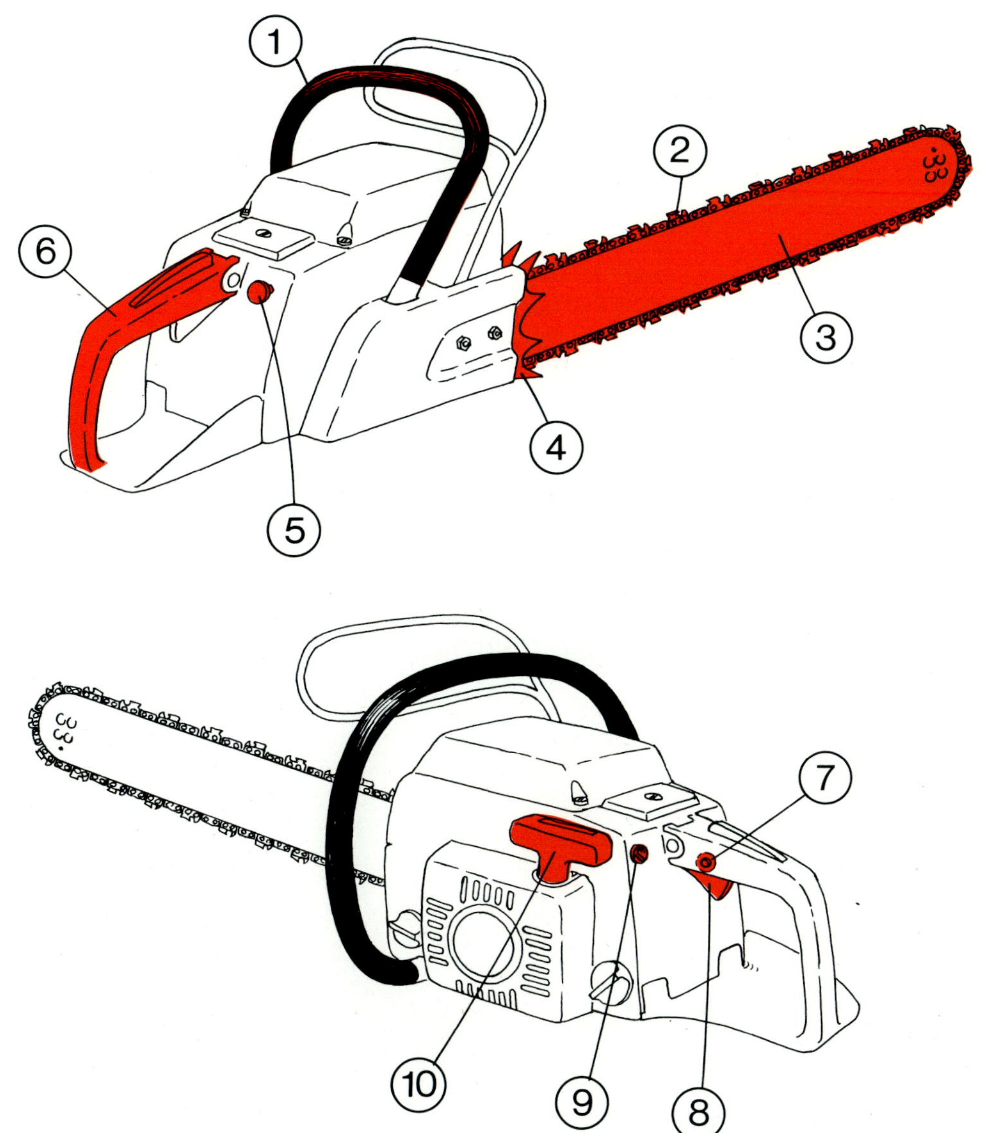
1
2
3
4
5
6
7
8
9
10

2.2 SAFETY DEVICES

Working with a chainsaw can be dangerous. Modern chainsaws therefore have several special safety devices. Chainsaws without the following safety devices should not be used:

1. Front handle guard with chain brake
   - protects left hand and stops saw chain during kickback.

2. Chain catcher
   - catches the saw chain if it breaks.

3. Rear handle guard
   - protects right hand.

4. Throttle control lock out
   - prevents saw chain from starting to move unexpectedly.

5. Anti-vibration devices
   - prevent vibration diseases of the hands.

6. Guide bar cover
   - avoids injuries during transport of chainsaw.

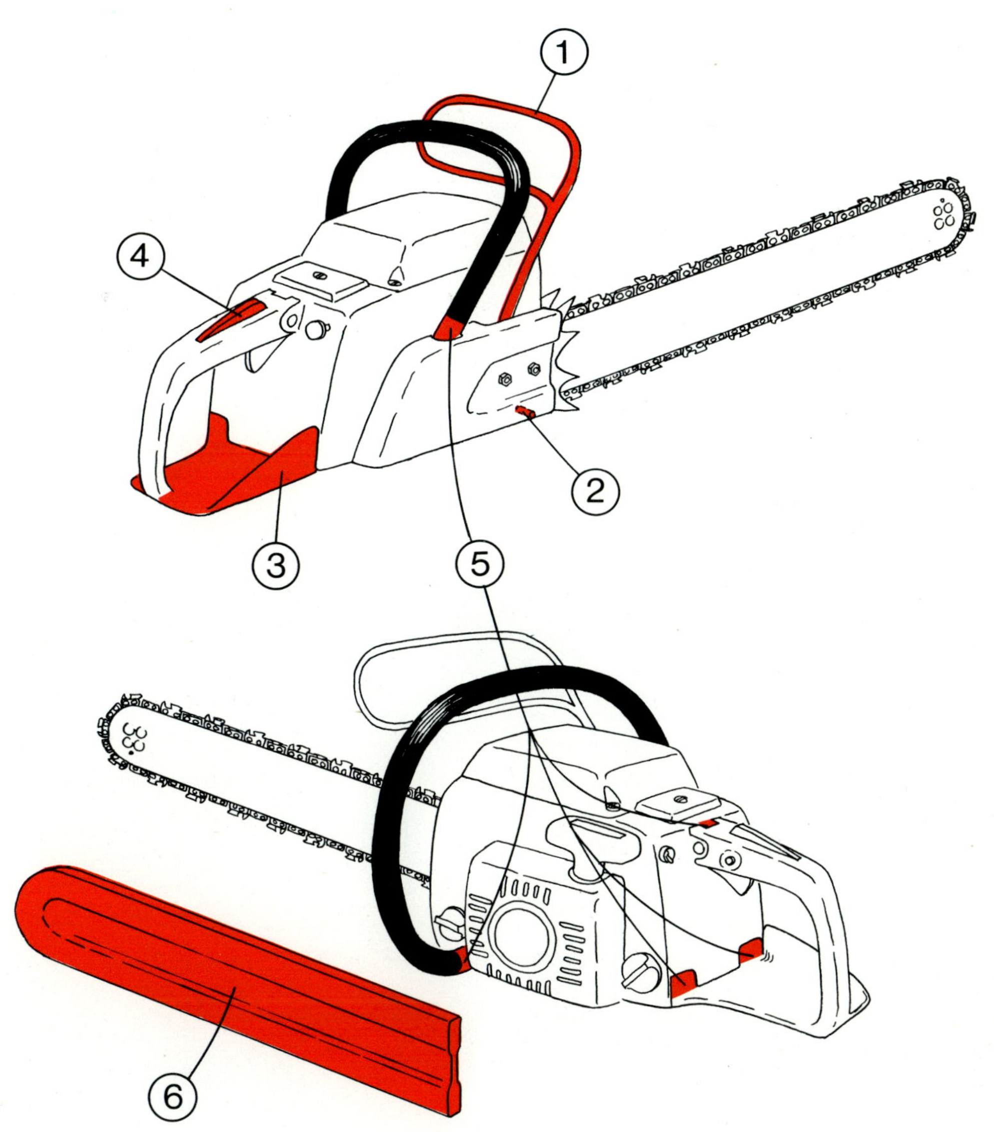
1
4
2
3
5
6

3. ACCESSORIES

3.1 FUEL AND CHAIN OIL CONTAINER, TOOL KIT

It is very practical to use a special container to carry fuel (1) and chain oil (2) in the forest. On 5 litres of fuel and 2 litres of chain oil the chainsaw will run from half a working day to a whole working day, depending on the cutting time. If necessary, the container can be refilled at the landing or shelter place during the meal break.

Always make sure that the fuel contains the right oil mix. The use of a funnel with a filter (3) helps to keep the tank clean and avoids spilling of fuel. Special chain oil is recommended for lubricating the chain. If not available, use engine oil or gear oil. Never lubricate the chain with waste oil since it rapidly destroys the oil pump. When refilling the fuel tank always also fill up the chain oil tank. First refill the oil tank, then the fuel tank.

A chainsaw tool kit must always be readily available in the forest. It should contain a T-wrench (4), a round file (5), a clamp (6), a spare chain (7), a spare air filter (8), and a small brush (9).

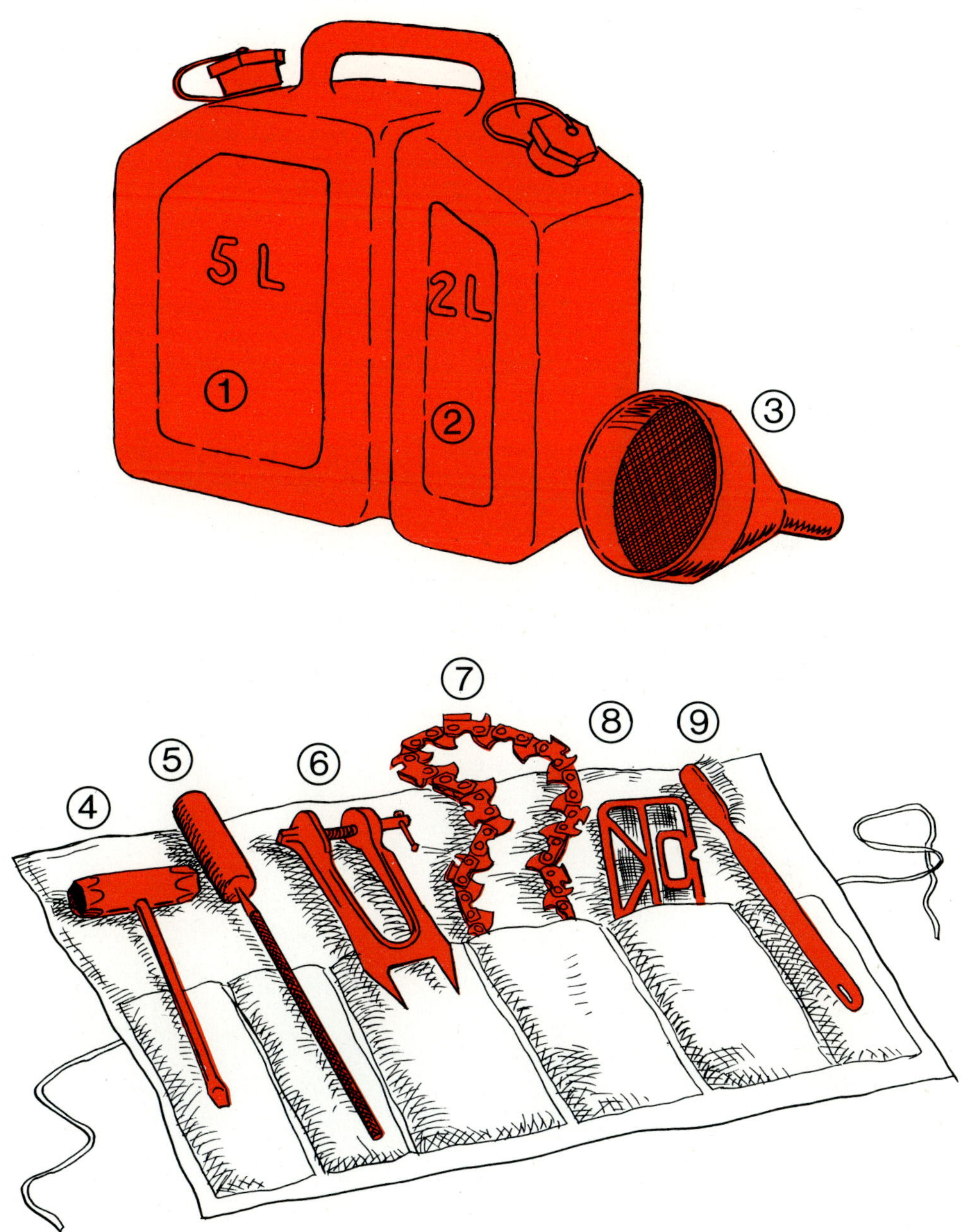
5 L
2L
①
②
③
④
⑤
⑥
⑦
⑧
⑨

3.2 ADDITIONAL EQUIPMENT

The following tools are needed in order to work efficiently with the chainsaw:

1. Matchets for clearing work, with protective cover, to be attached to the workers' belt.

2. Wedges for keeping the saw cut open and directing the direction of fall of the tree. Wedges should be made of soft metal or wood. Steel must not be used because it can destroy the saw chain and splinters can injure the operator.

3. An Axe for clearing work, for delimbing of small trees and for driving wedges. Weight of axe should be about 1 000 grams, length of handle 70–80 centimeters, with protective cover.

4. A Hammer for driving wedges in bigger trees. Weight about 3 000 g, handle straight and about 80 cm long.

5. A Cant Hook with pole for taking down or turning medium-sized trees.

6. Tape Measure or Scale Stick and Caliper (as required).

For small sized trees the following special equipment is recommended:

7. A Felling Lever with cant hook.

8. Hooks or Tongs for wood handling.

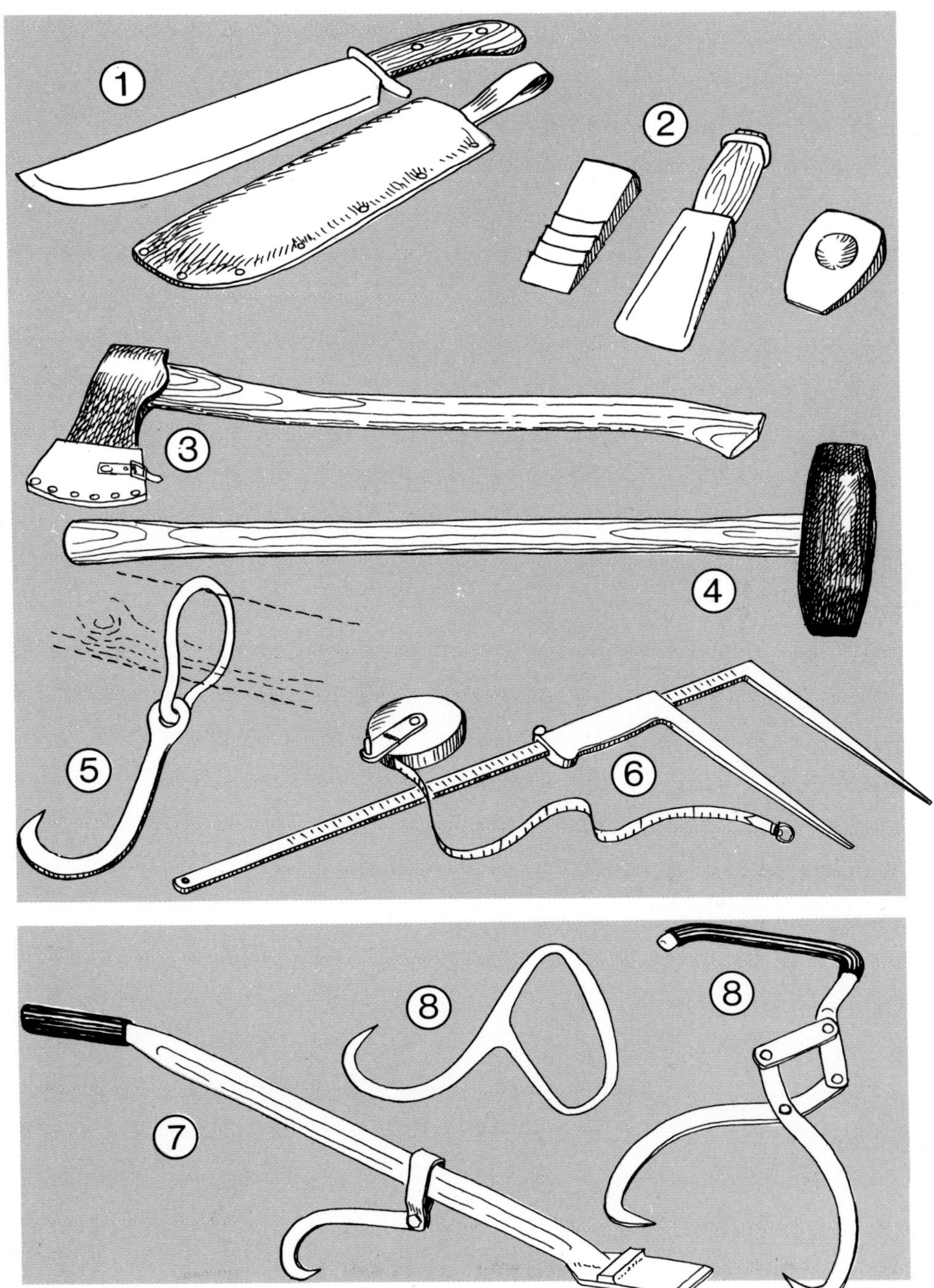
1
2
3
4
5
6
7
8
8

4. THE OPERATOR

4.1 CLOTHING AND PERSONAL PROTECTIVE EQUIPMENT

The chainsaw operator and his helper(s) should wear:

1. Long-sleeved shirt or jacket, preferably in a warning colour, fitting neither too loose nor too tight.

2. Long trousers.

3. Boots with non-slip soles.

The chainsaw operator and his helper(s) must wear:

4. Safety helmet which has ventilation holes. The helmet provides protection against falling branches and other material.

The operator's helmet should be fitted with:

5. An eye protector, as a protection against sawdust and other flying particles.

6. Ear protectors are advisable for the operator if he works regularly with the chainsaw for several hours a day over a period of several years.

7. Gloves serve as a very effective protection against splinters, cuts, bruises, burns and dirt.

8. A pocket first aid kit should contain roller bandages and standard dressings for open wounds.

A full first aid kit should always be available nearby, for treating serious accidents.

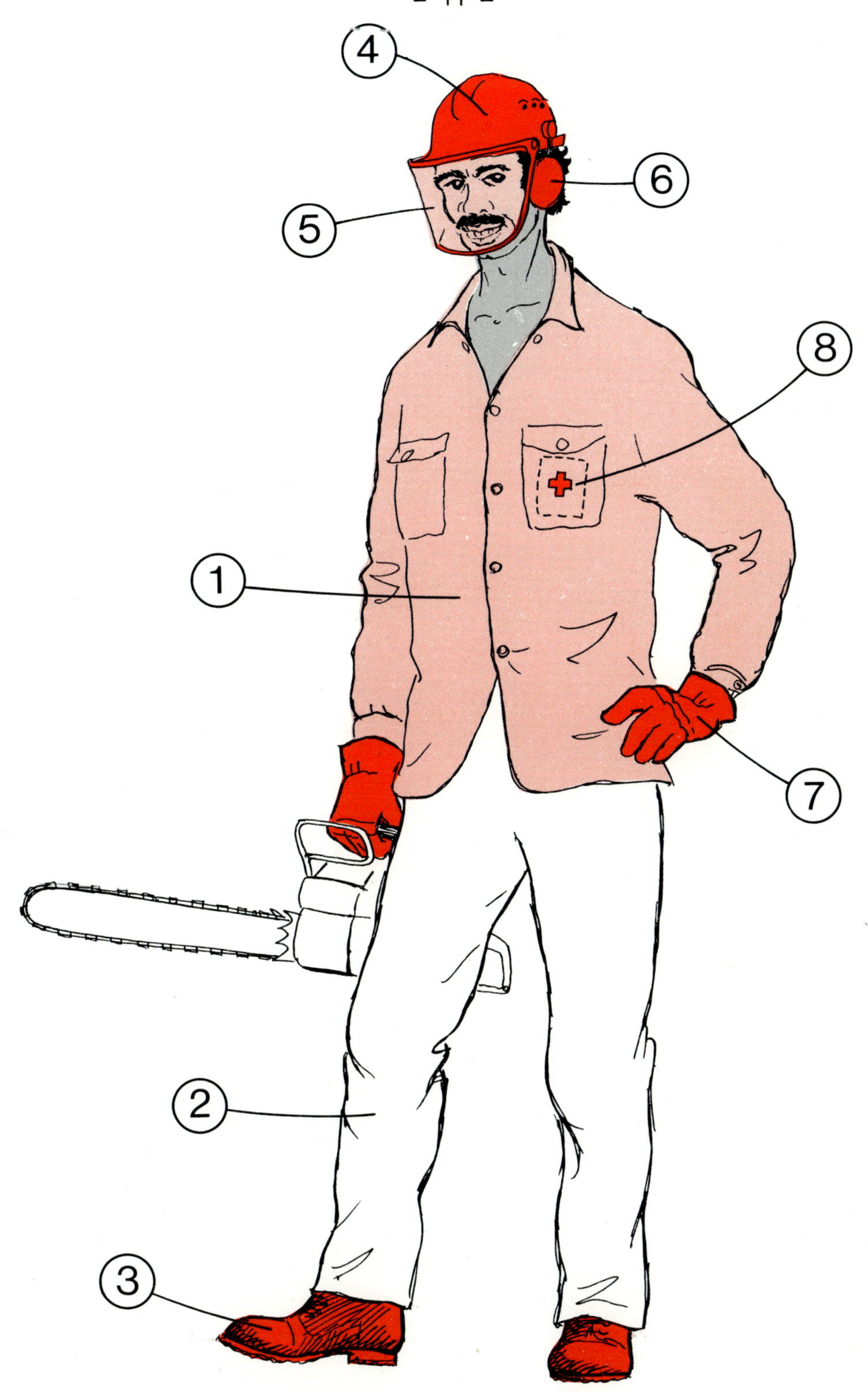
4
6
5
8
1
7
2
3

4.2 FOOD AND NUTRITION

Work with a chainsaw is heavy and over a period of time is tiring. The operator must therefore be in a good state of health and well fed. Food should be taken before starting work and during meal breaks. Under severe conditions, not more than six hours of productive work can be expected. In this case two meal breaks of at least 30 minutes each should be provided after the first 2 hours of work and after the following 2 hours of work.

Food should be sufficiently rich in starches (rice, maize, wheat, millet, banana, cassava); in protein (eggs, meat, fish); in fats (olive oil, coconut oil, cheese, butter); and in vitamins (fruit, vegatables).

During hard work and in hot weather the body can lose 3 to 6 litres of liquid per day. This must be replaced. Workers should carry a container with boiled water, tea or other beverages and drink when they are thirsty, or at regular intervals.

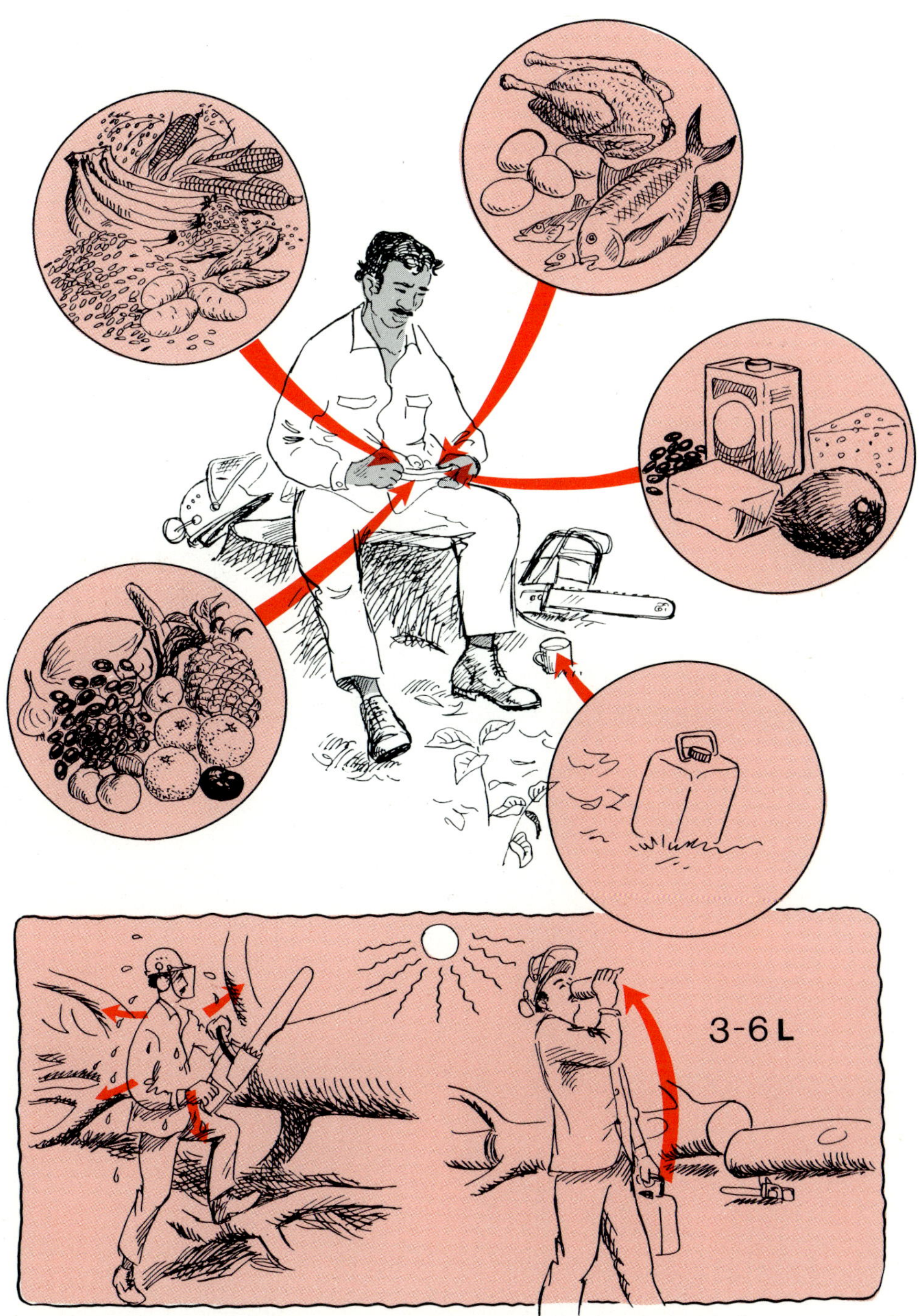
3-6 L

5. BASIC HANDLING OF THE CHAINSAW

5.1 STARTING THE CHAINSAW

1. Take the chainsaw away from the place where it was filled, in order to avoid starting a fire.

   To avoid kickbacks make sure that there are no obstacles near the chain.

   If the engine is cold set the choke and the throttle control latch.

   If possible, place the right foot in the rear handle. Grasp the front handle firmly with the left hand, placing the thumb around the bar.
   With the right hand give a short, sharp pull when the starter mechanism is engaged. Hold the starter handle as the rope rolls back.

   When the engine is running, release the throttle control latch and choke.

2. Check whether the chain is receiving sufficient oil. Never start the engine when the guide bar and chain are dismantled because the engine may break down. Stop the motor when carrying the chainsaw, except for short distances from cut to cut.

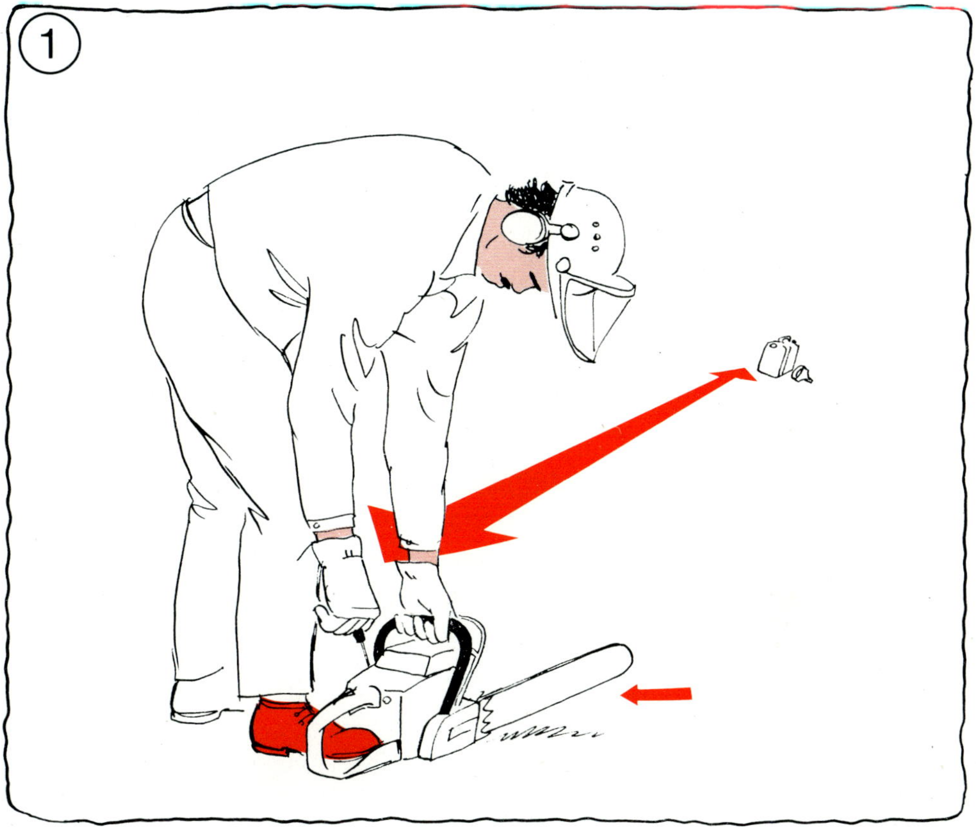

1

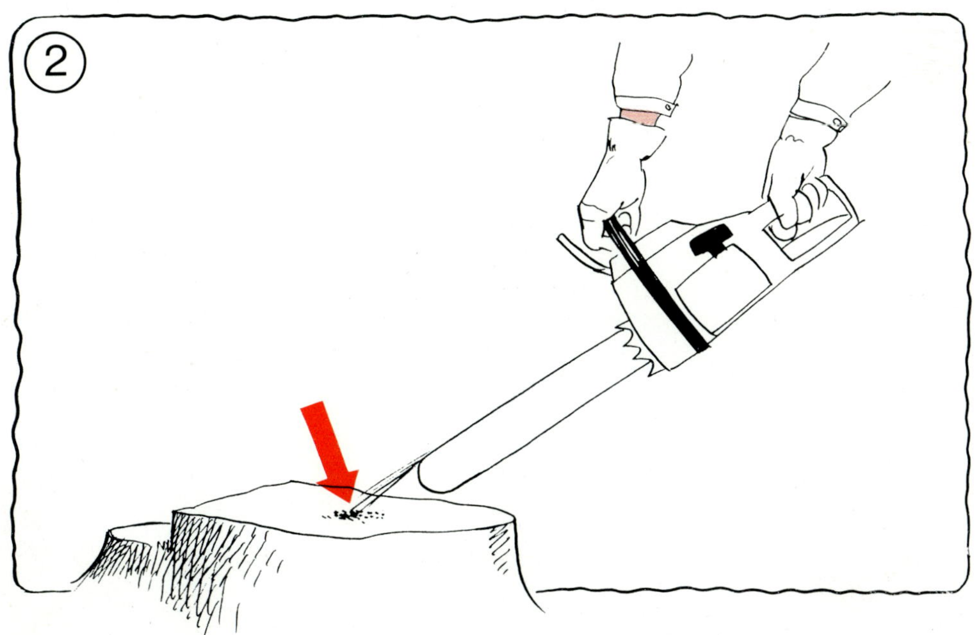

5.2 BASIC RULES IN CHAINSAW HANDLING AND CUTTING

1. Stand with your legs well apart, one foot forward, firmly on the ground, and ensure that you will not slip.

2. The left thumb must always be placed around the front handle in order to keep a tight grip.

3. Keep the chainsaw close to the body and try to support its weight on the body or on the tree.

4. Keep a distance of at least 2 metres from other persons when the saw chain is running.

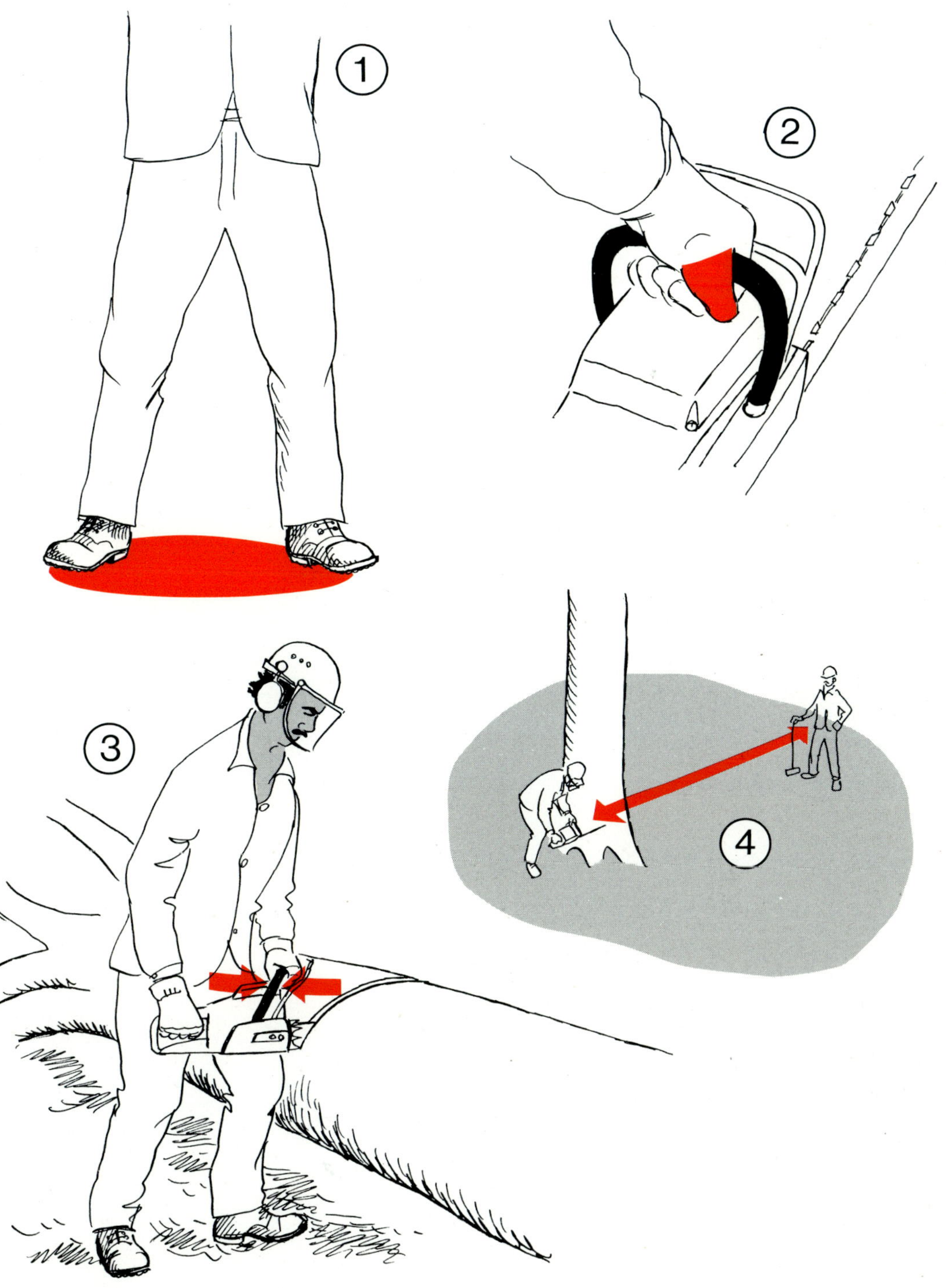

1
2
3
4

5.2 Basic Rules in Chainsaw Handling and Cutting (Continued)

Always use the shortest possible guide bar. This facilitates handling of the chainsaw and reduces the risk of accidents.

1. Cutting with the backward-running chain is the safest and easiest technique. The action of the chain pulls the saw towards the tree.

2. Cutting with the forward-running chain requires more effort because it pushes the saw away from the cut and towards the operator. This technique should only be used when cutting with the backward-running chain is not practical.

3. Cutting with the tip of the saw must be avoided because it can cause kickbacks. A kickback is a sudden upward and backward movement of the guide bar and is extremely dangerous. Kickbacks can also occur when the running chain touches a branch or log, or when it is pinched in the cut.

4. If a boring cut is made, first cut with the side of the guide bar, then carefully lift the saw up.

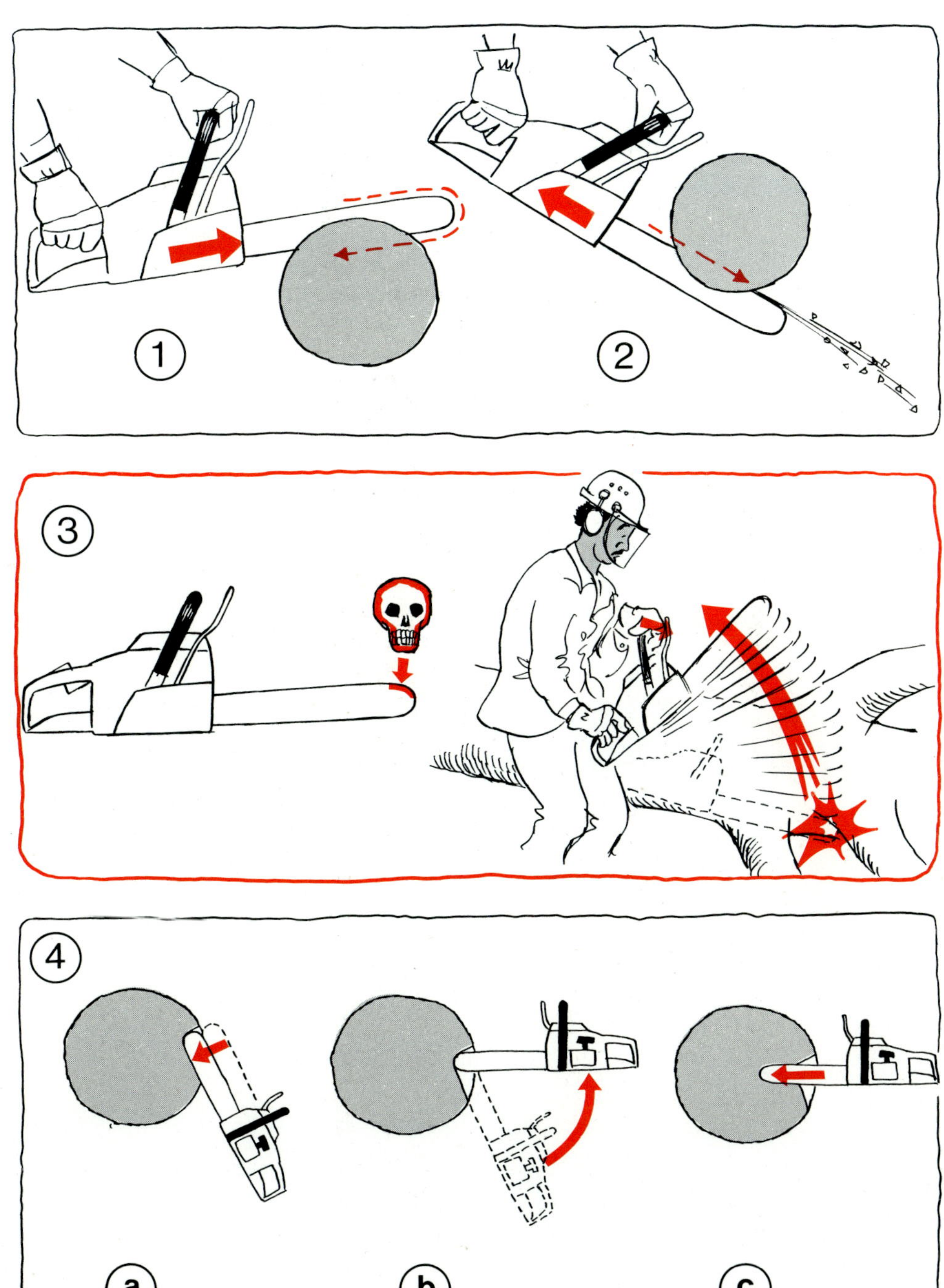
1
2
3
4
a
b
c

6. BASIC TREE FELLING

6.1 PREPARATION FOR TREE FELLING

Felling is the most dangerous job in forest operations. It requires skilled men and a carefully planned working routine.

1. The felling team must keep a minimum distance of two tree lengths from other workers. This distance may be increased to four tree lengths when visability in the forest is poor.

   The felling direction must be carefully determined. This will depend on the skidding direction, the lean of the tree, the wind, on obstacles in the way of the tree's fall and on obstacles on the ground, and also on the possibility of retreating safely.

2. When the felling direction (a) is determined the tools are placed opposite to the felling direction, behind the tree (b).
   The working area around the tree is cleared (c).
   Two escape routes (d) are cleared, as far as necessary, leading sideways at about 45 degree angles to the rear.

3. The base of the tree must be well cleared, using matchets or an axe, in order to prevent the saw chain becoming blunt too quickly.

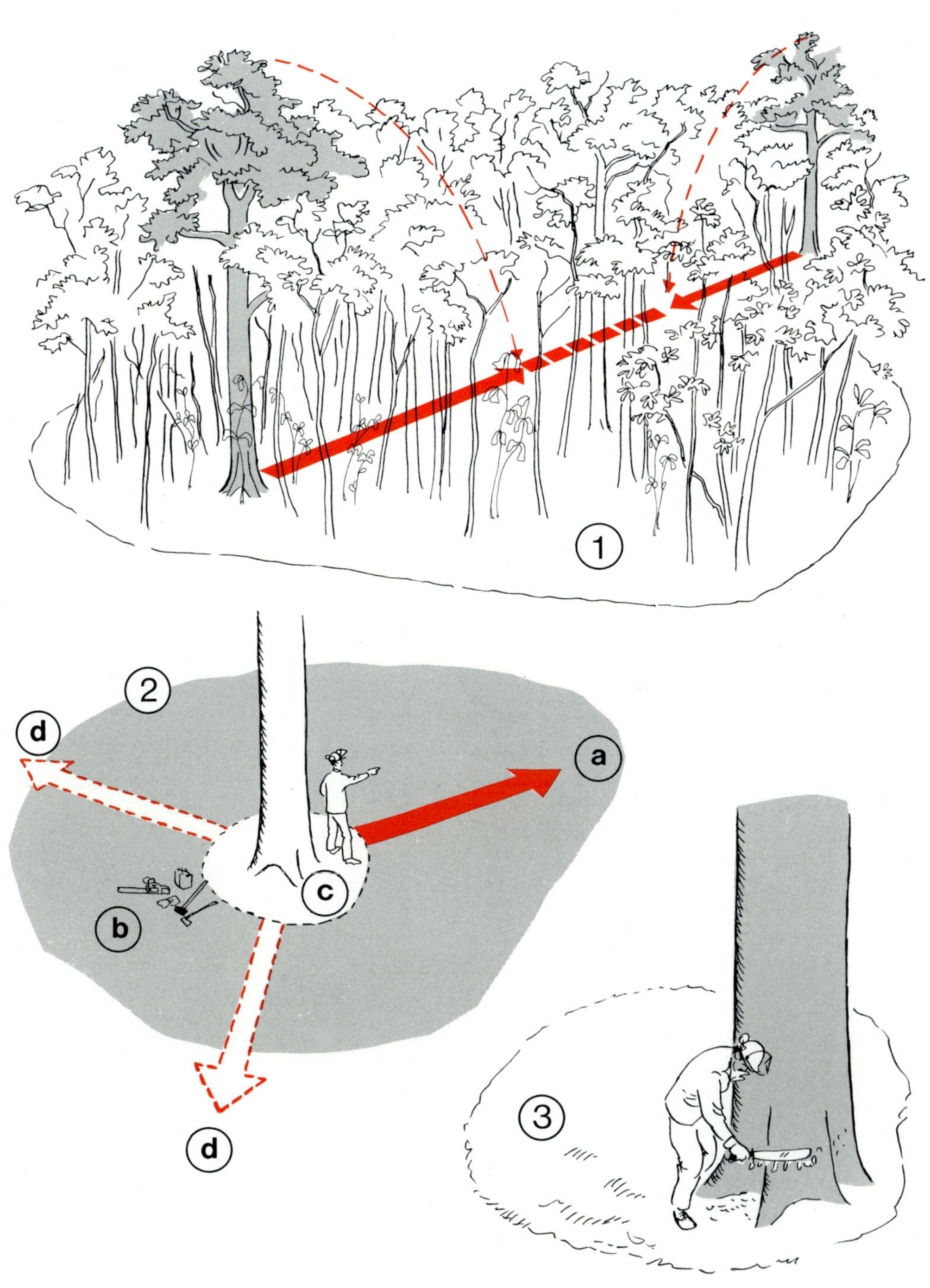
1
2
d
a
c
b
d
3

6.2 FELLING OF SMALL TREES

Small trees, such as in thinnings, are usually felled by one chainsaw operator. If delimbing is to be done by axe the team may consist of two or more men. In such cases it is an advantage if more than one man knows how to handle the chainsaw.

Trees up to approximately 60 cm diameter at the stump are easily felled in the desired direction if they are regular in shape.

After determining the felling direction (1) and clearing the tree's base and escape routes, sawing starts with the undercut (2) which should penetrate about one-fifth to one-quarter of the diameter into the tree. The undercut should be made close to the ground if the base of the tree is to be utilized or if high stumps would obstruct subsequent operations, such as skidding. The opening of the undercut should be at an angle of about 45 degrees.

The oblique cut (3) is made first. The horizontal cut (4) must meet the oblique cut in a straight line facing the felling direction at an angle of 90 degrees. If this is not the case, the undercut must be corrected. Proper and careful work in the undercut phase avoids the manifold troubles which may arise from uncontrolled and careless work.

If stumps are liable to tear splinters from the tree, as is often the case with softer woods, the undercut should be terminated by small lateral cuts (5) on both sides of the hinge (6).

The back cut (7) must also be horizontal. It should be placed about 2.5 to 5 cm higher than the base of the undercut. The hinge is necessary to guide the tree during its fall. It should have the same width on both sides.

If the tree's diameter is smaller than the guide bar, the back cut can be made in one movement (8).

If it is larger, the saw must be shifted several times (9).

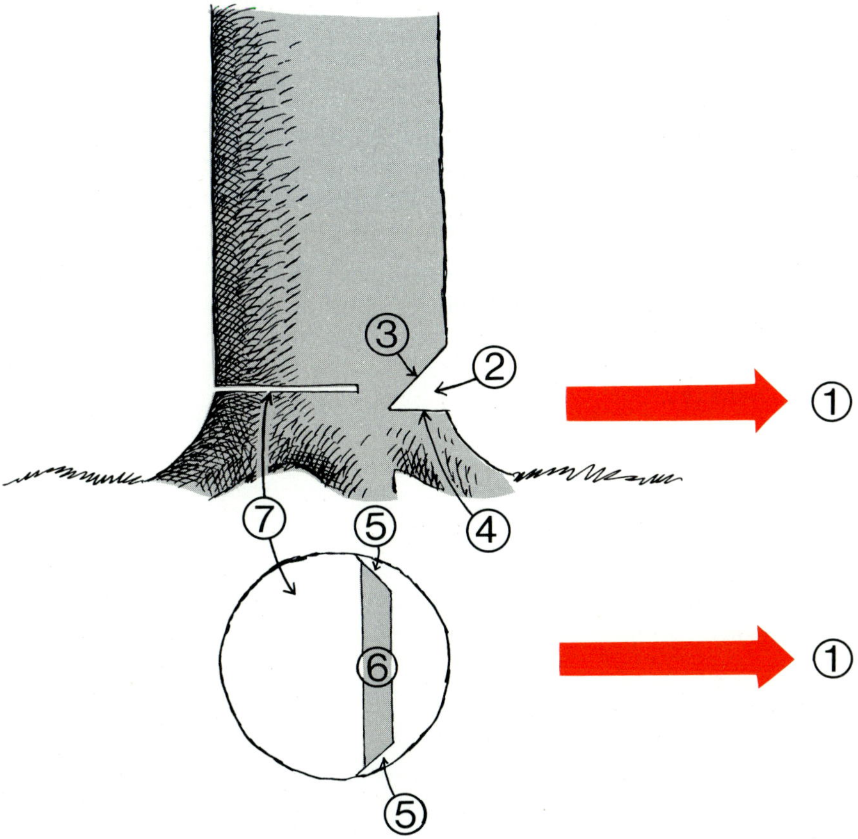

③ ② ①

⑦ ④

⑤

⑥ ①

⑤

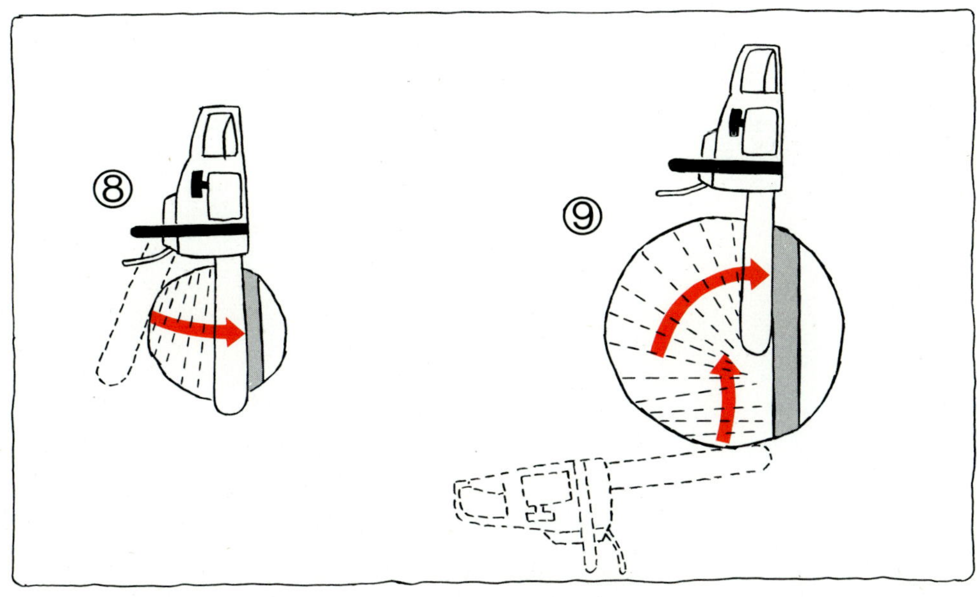

6.3 PRECAUTIONS WHEN MAKING THE UNDERCUT AND THE BACKCUT

Accurate felling makes the job safer, facilitates subsequent operations and reduces timber wastage. Felling should therefore be done with the greatest care and precision. By looking at the stump one can easily see whether a poor or a good felling job has been done.

The undercut is the most important step in felling.

1. When making the undercut, care must be taken that it points precisely into the felling direction. This can be checked by looking along the line of the front handle.

2. The oblique cut (a); the horizontal cut (b); and the backcut (c) must not be carried too far, as shown. Sufficient holding wood which acts as a hinge, must always remain in order to maintain control of the tree so that it does not break, slip, or twist off the stump and fall in any direction other than that intended.

3. Before starting the backcut the operator must give a loud warning shout.

4. Important aids in felling are a wedge (a), or a felling lever (b), to push the tree over into the felling direction when the backcut has been terminated, if necessary.
   Once the tree starts falling, the operator must retreat backwards along one of the escape routes and watch out for falling branches and other debris.

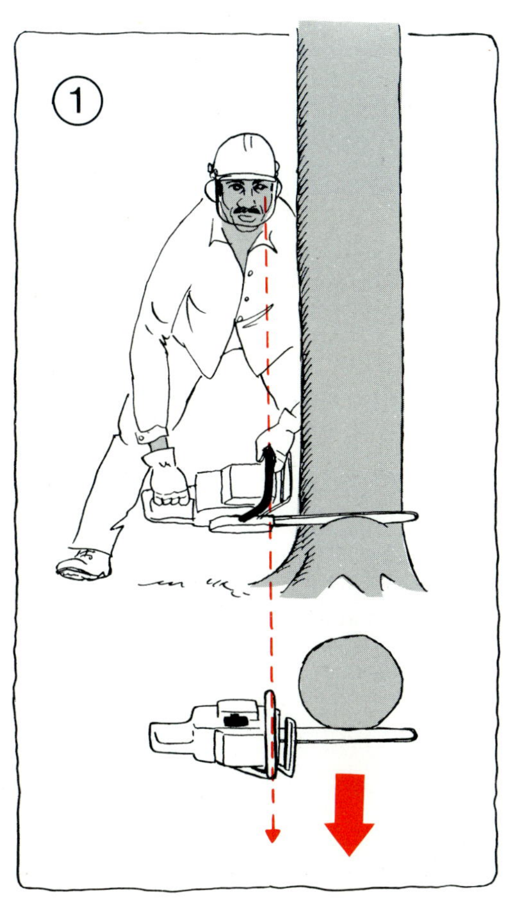
1

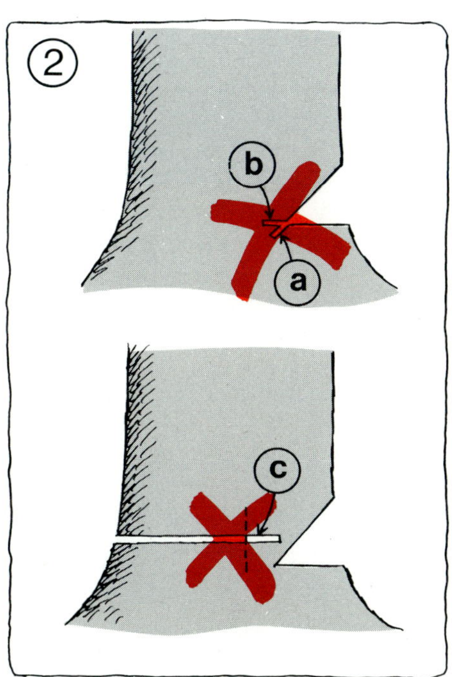
2
b
a
c

3

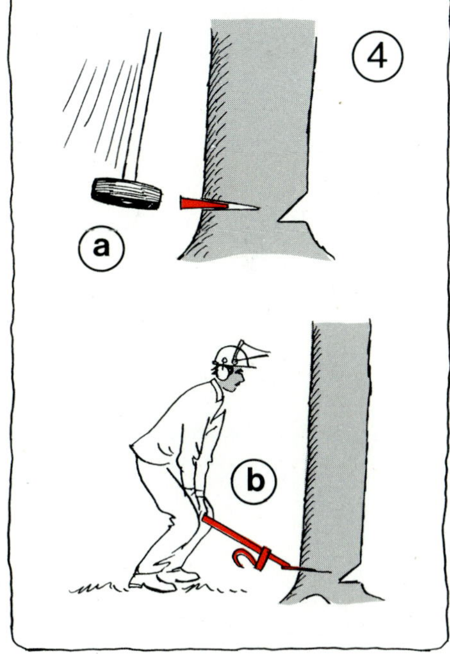
4
a
b

6.4 FELLING OF LARGE TREES

1. If the tree's diameter is up to twice the length of the guide bar, it may be necessary to saw the undercut from both sides. To avoid pinching, the horizontal cut of the undercut should be made first, followed by the oblique cut.

2. Next a centre cut is made from the side of the notch of the undercut (a).
   Finally, the backcut is made (b), while retaining a hinge on both sides of the tree of at least 5 cm thickness (c). The backcut should be 10-20 cm (or more) higher than the base of the undercut (d).

3. Small buttresses should preferably not be removed in advance because it is safer to leave them on.
   If this is required to facilitate skidding, loading and conversion, this can usually be done more conveniently on the felled tree (a). If the guide bar is too short, it may, however, be necessary to remove one or even all buttresses before felling (b).
   As a rule it is preferable to use a rather short guide bar in order to carry less weight and to make handling of the chainsaw easier.

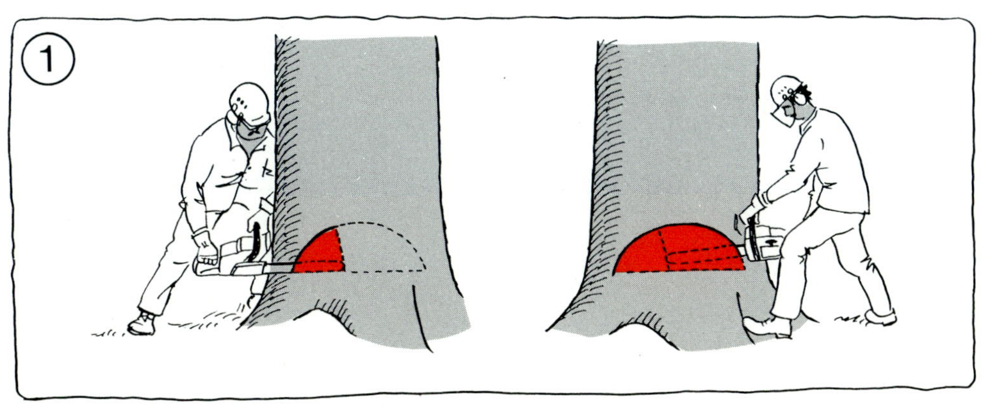
1

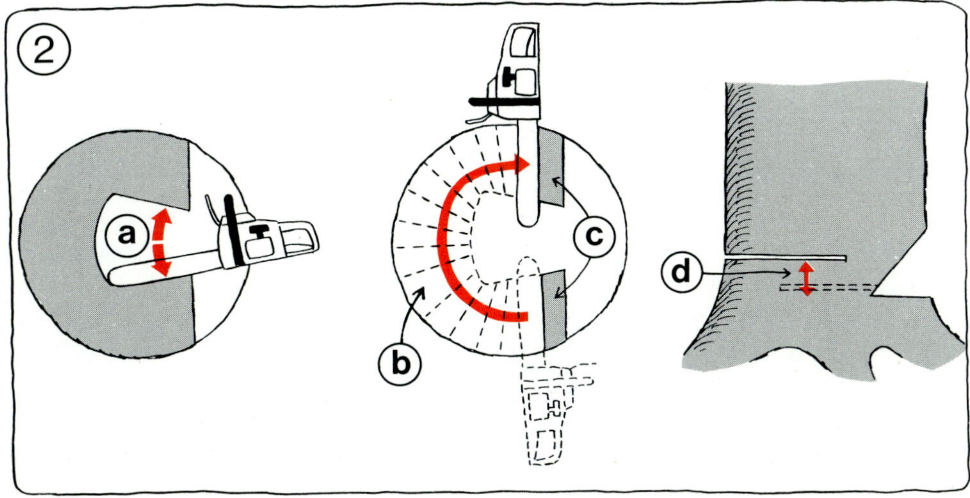
2
a
c
d
b

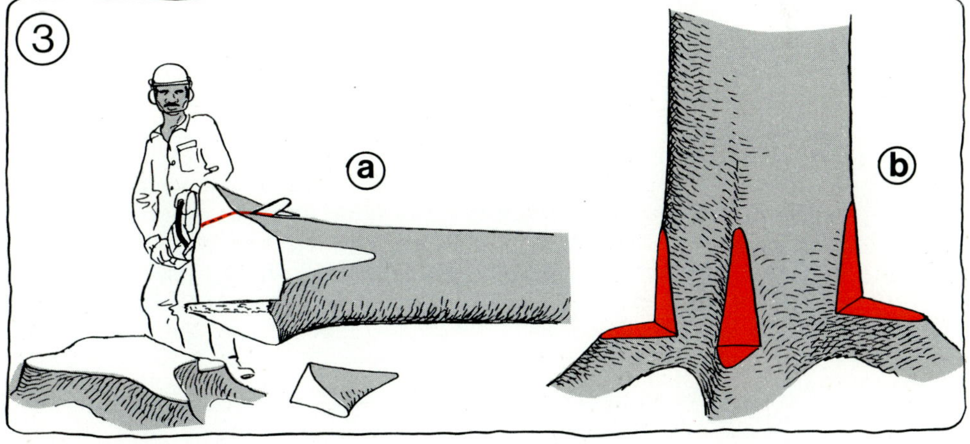
3
a
b

6.5 FELLING OF LEANING TREES

If the tree leans heavily into the direction of fall the following technique helps to avoid cracking of the tree or pinching of the chainsaw.

1. On a small tree after making the undercut (a), the backcut is carried out in three sections. The two sides (b) are cut first, followed by the remaining part (c).

2. On a larger tree the undercut (a) should not exceed about one-quarter of the diameter. Otherwise the guide bar might get pinched. The backcut is then started by a boring cut from one side (b). In larger trees a second boring cut is made from the other side (c). Sufficient anchorage wood (d) is left behind. This is finally sawn with an oblique cut.

3. A tree can also be felled at an angle of about 30 degrees from the lean. In this case the undercut (a) should face the intended felling direction (b). The hinge (c) should be kept smaller on the side of the lean (d) and larger on the side to which the tree is to be felled. In addition, a wedge (e) placed on the side of the lean will help to direct the tree's fall.

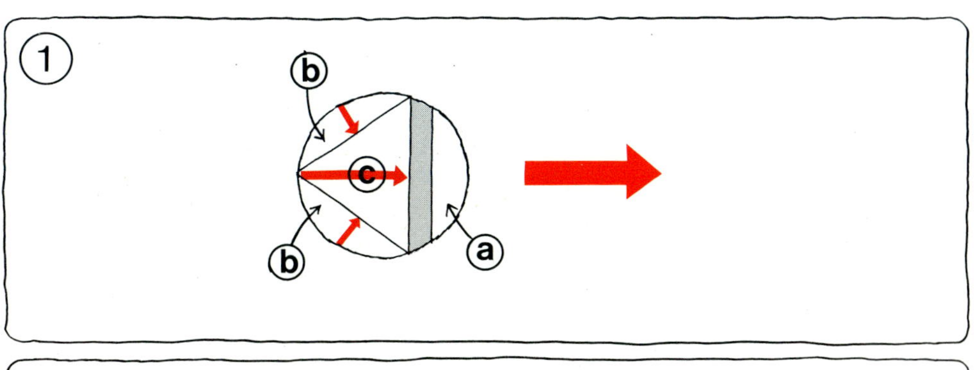
1
b
c
b
a

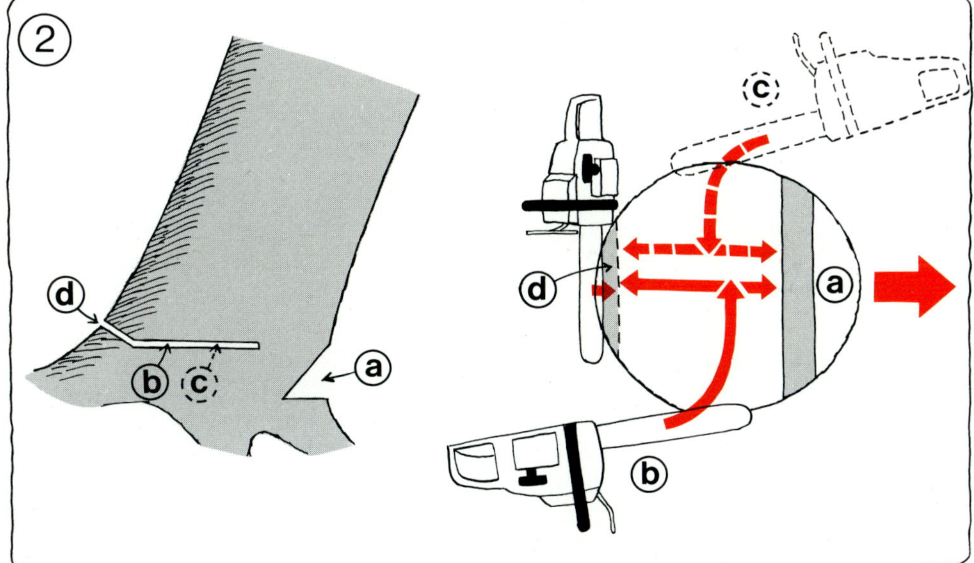
2
d
b
c
a
c
d
a
b

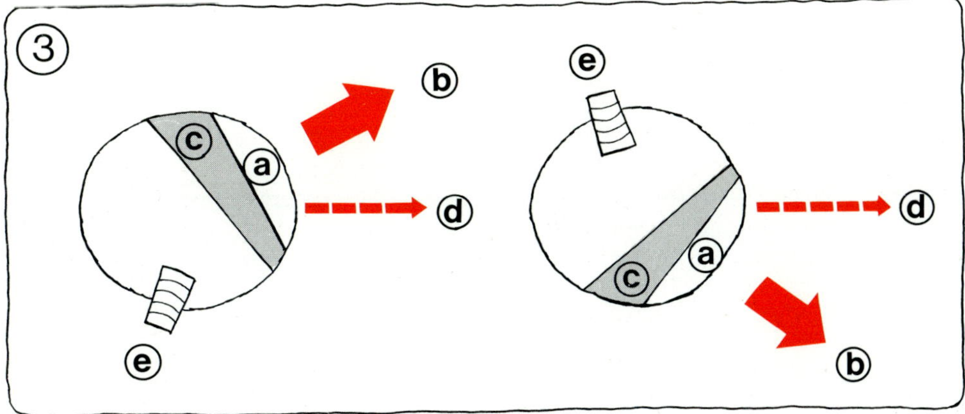
3
b
c
a
d
e
e
d
a
c
b

7. FELLING OF LARGE TROPICAL TREES

7.1 FIELD PLANNING IN DENSE NATURAL FORESTS

Natural forests in tropical regions differ greatly from man-made forests. Trees suitable for utilization are irregularly distributed and volume and number per hectare is low. This calls for special advance planning to facilitate logging operations, to make them more efficient and to make the work easier and safer.

1. Stock Map

   A stock map shows the approximate position of individual trees to be felled It also includes information on terrain difficulties such as rivers, creeks, swamps and rocks. Windfalls (trees blown down by the wind) may also be recorded. The stock map is prepared during timber cruising (or enumeration). It is divided into squares based on the cruising strips in the field.

2. Operation Map

   An operation map is based on the stock map. In addition it shows roads (a), landings (b), bridges (c), and skid trails (d). (See below). The operation map is very useful for field personnel. Every felling team should have its own copy. A handy map size may be at a scale of 1 : 5 000 and should cover an area of about 200 m x 200 m = 4 ha.

(a) Roads

(b) Landings

(c) Bridges

(d) Skid trails

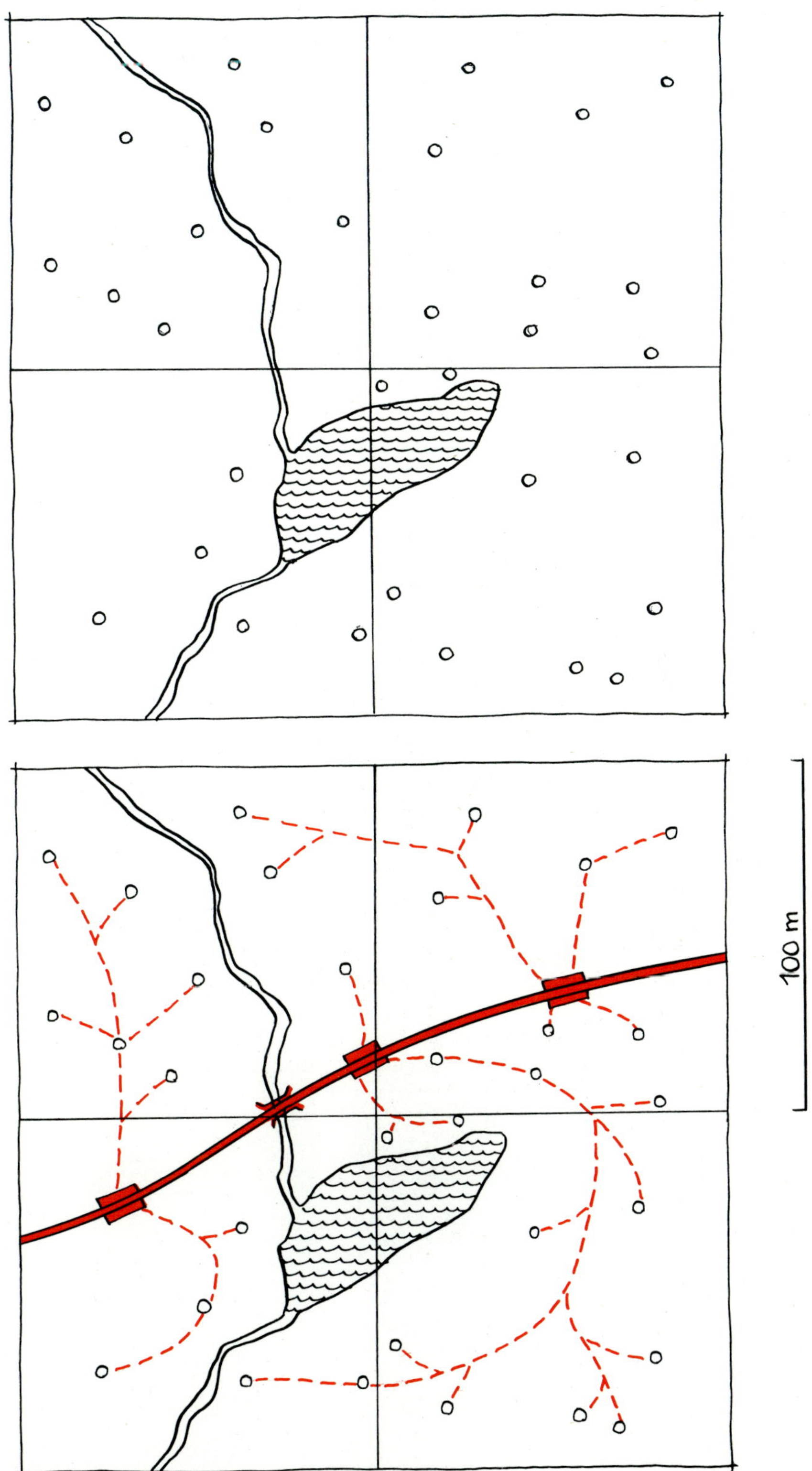
100 m

7.2 THE FELLING TEAM AND ITS EQUIPMENT

Usually the felling team consists of three men:

(1) The Operator - who should be well trained in sawing techniques and basic chainsaw maintenance;

(2) The Assistant - who guides the operator, drives wedges and helps in clearing work. He should also be able to saw when the operator is tired;

(3) The Helper - who is used for clearing work, for carrying tools and for filling up the chainsaw with fuel and chain oil.

If the stock of harvestable trees is dense the team is usually reduced to two men. In this case the assistant does the helpers job.

The workers should always remain in one team and keep the same equipment. The equipment may be divided among the team as follows:

(1) The Chainsaw Operator
The chainsaw with a protective chain cover (6-12 horsepower; guide bar 45-80 cm); the tool kit, a matchet with a protective cover, a safety helmet with eye protector and possibly ear protectors, and a pocket first aid kit.

(2) The Assistant
The hammer, the wedges, an axe, a sharpening stone, a measuring tape, a matchet with protective cover, a safety helmet with eye protector, and possibly ear protectors and a pocket first aid kit.

(3) The Helper
The fuel/oil container, the funnel, a water can, a matchet with protective cover, a safety helmet and a pocket first aid kit.

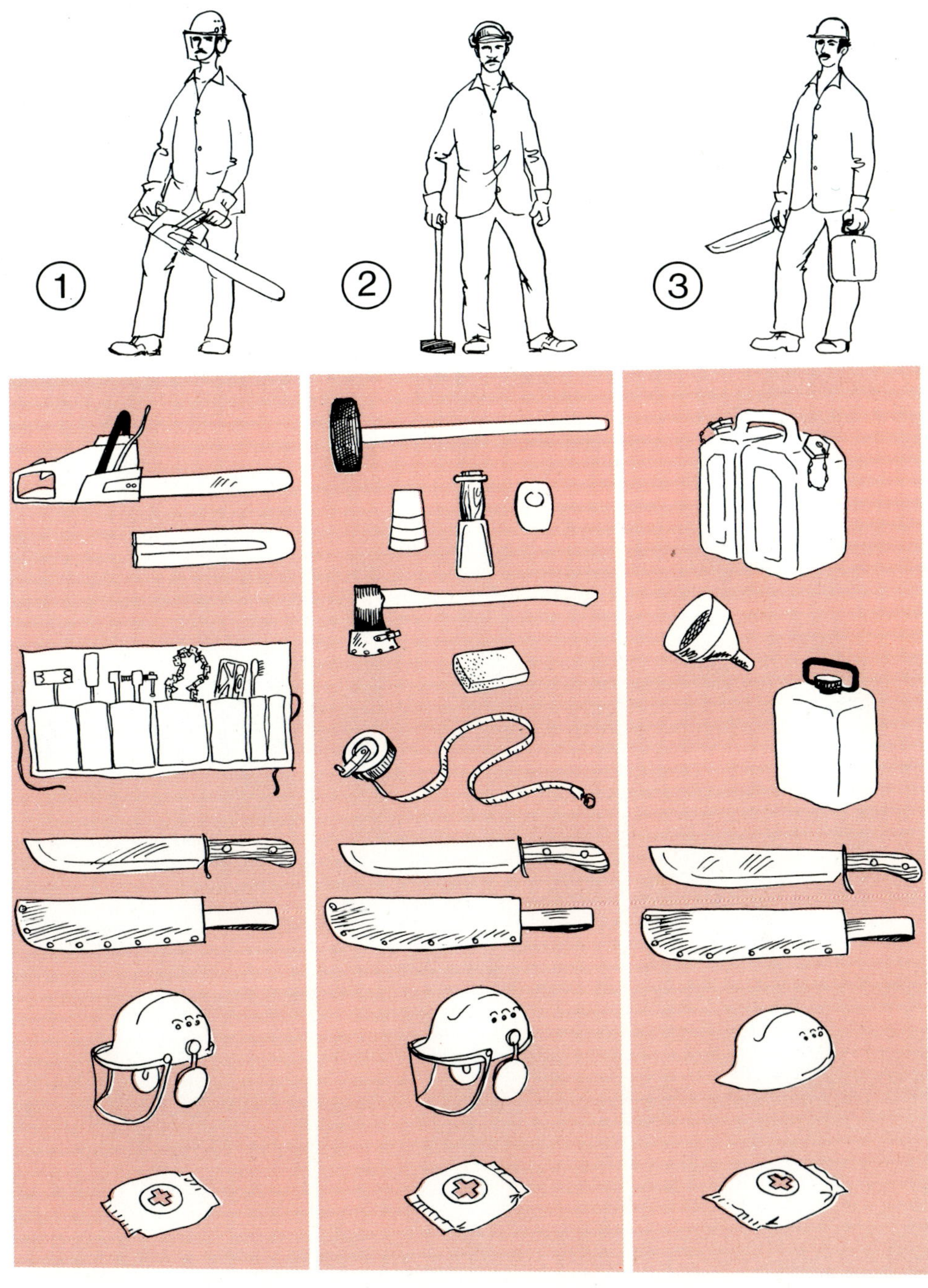
1
2
3

7.3 SPECIAL ACCIDENT RISKS IN DENSE TROPICAL FORESTS

Felling in dense tropical forests is particularly dangerous. Accident rates can be extremely high if the specific risks are not respected.

Dense undergrowth makes it difficult to retreat from the tree during its fall. Dead branches may be hidden in the crown which is often not visible. For the same reason it may be difficult to assess the lean of the tree.

Trees may be over-mature and may therefore have hollow or rotten centres.

Trees are often connected to each other with climbers. When the trees fall they frequently pull down other trees (1). Branches from the falling tree or from neighbouring trees (2)(3) are broken off and may swing backwards (4). Climbers are torn off or may break and snap back (5).

1
2
3
4
5

7.4 SPECIAL PRECAUTIONS WHEN PREPARING FELLING WORK

The accident risk when felling trees in dense tropical forests is considerably reduced if the area around the base of the tree and the escape routes are well cleared.

Two paths are cleared at a length of 20-30 m beyond the reach of the crown opposite to the felling direction (1). The angle between them should be 45 degrees or more.

Clearing work is done by the helper with his matchet (2). The assistant (3) joins him in this work as long as he is not needed to guide the operator (4), or to drive wedges.

Before starting to saw, the operator checks the base of the tree and removes dirt and loose bark at the places where the saw cut will be made (5).

Climbers attached to the tree (6) must also be cut before sawing begins.

1
2
1
6
5
4
3

7.5 BUILDING OF PLATFORMS

Additional difficulties may exist in steep terrain where the operator may not be able to stand safely on the ground, and it may be impossible to use the proper felling techniques because the undercut and the backcut cannot be made at the desired levels.

In such cases it may be necessary to construct a platform from material available locally. Experienced workers will be able to do this rather quickly. If the operator can stand on the platform the felling work is much easier and safer and the risk of wasting wood because of inadequate felling techniques is avoided.

There are different ways of constructing platforms depending on the shape of the tree, the terrain and the material available for building.

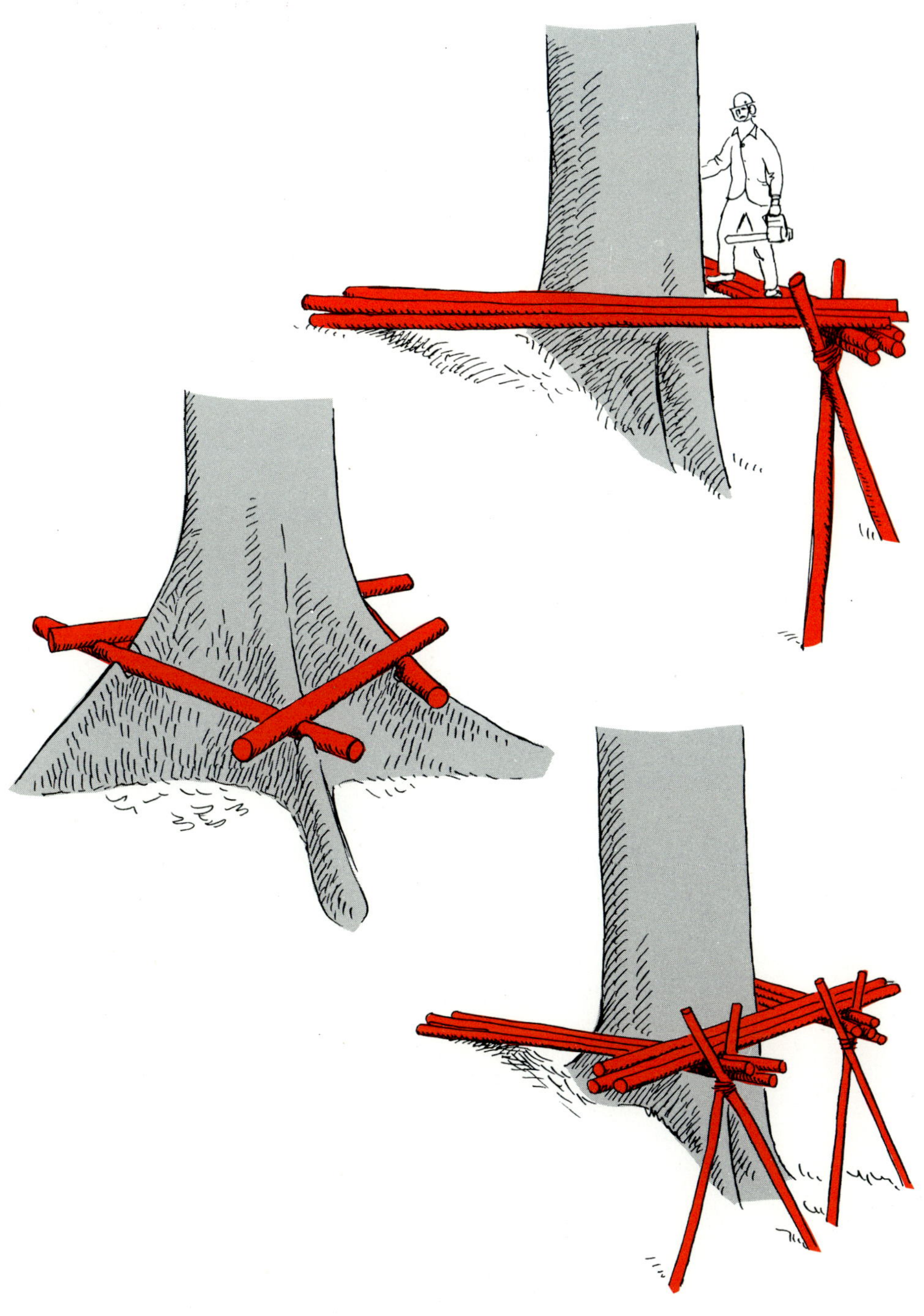

7.6 TREES WITH PLANK BUTTRESSES

Plank buttresses are common in tropical trees. They occur in many tree species once they have grown large.

Large trees with plank buttresses often attain a cylindrical shape only at a height of 3-5 m above ground level (1). At ground level the cross-section becomes larger and more irregular (2).

Trees with large plank buttresses are felled at a height of about 80 cm above ground level (3). At this height the tree will have sufficient centre wood to remain stable once the buttresses have been cut so that it can be felled in the desired direction.

If the buttress part of the log cannot be utilized it can be cut off after felling (4).

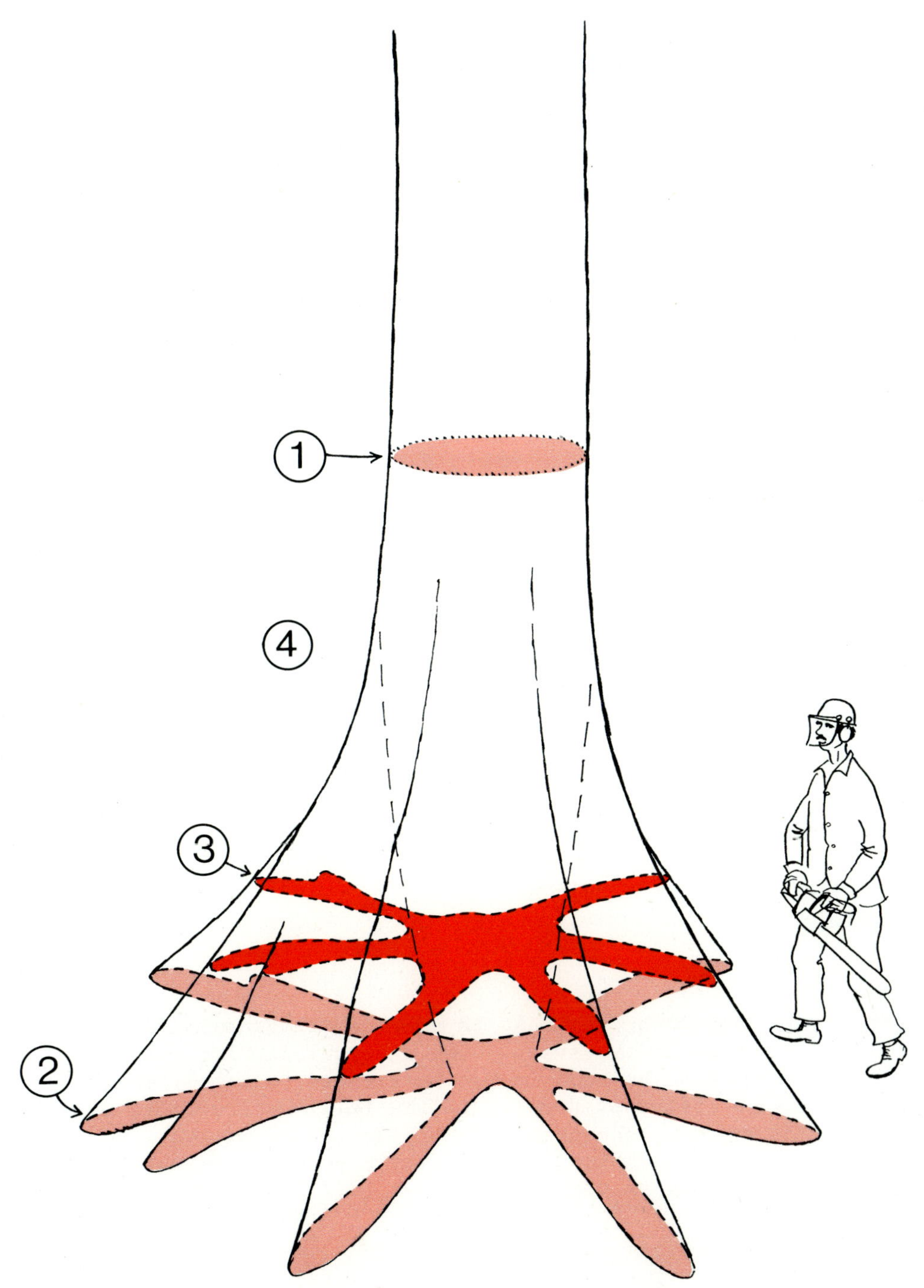
1
4
3
2

7.7 FELLING TREES WITH PLANK BUTTRESSES

1. The undercut (a) is made at a height of about 80 cm to a depth of about one-third of the diameter.

   First the horizontal cut is made and then the oblique cut.

   The backcut is made about 20 cm higher (or more) than the horizontal cut of the undercut. The cut begins on the side buttresses (b) and is finished on the rear buttress (c).

2. This technique must be adapted to the particular shape of the tree. The undercut may for instance have to be made in two buttresses (a). If there is no buttress opposite to the felling direction, as in the case shown in illustration 2., first cut (b) is made, then the lateral buttresses (c) are cut. Felling is completed by cut (d).

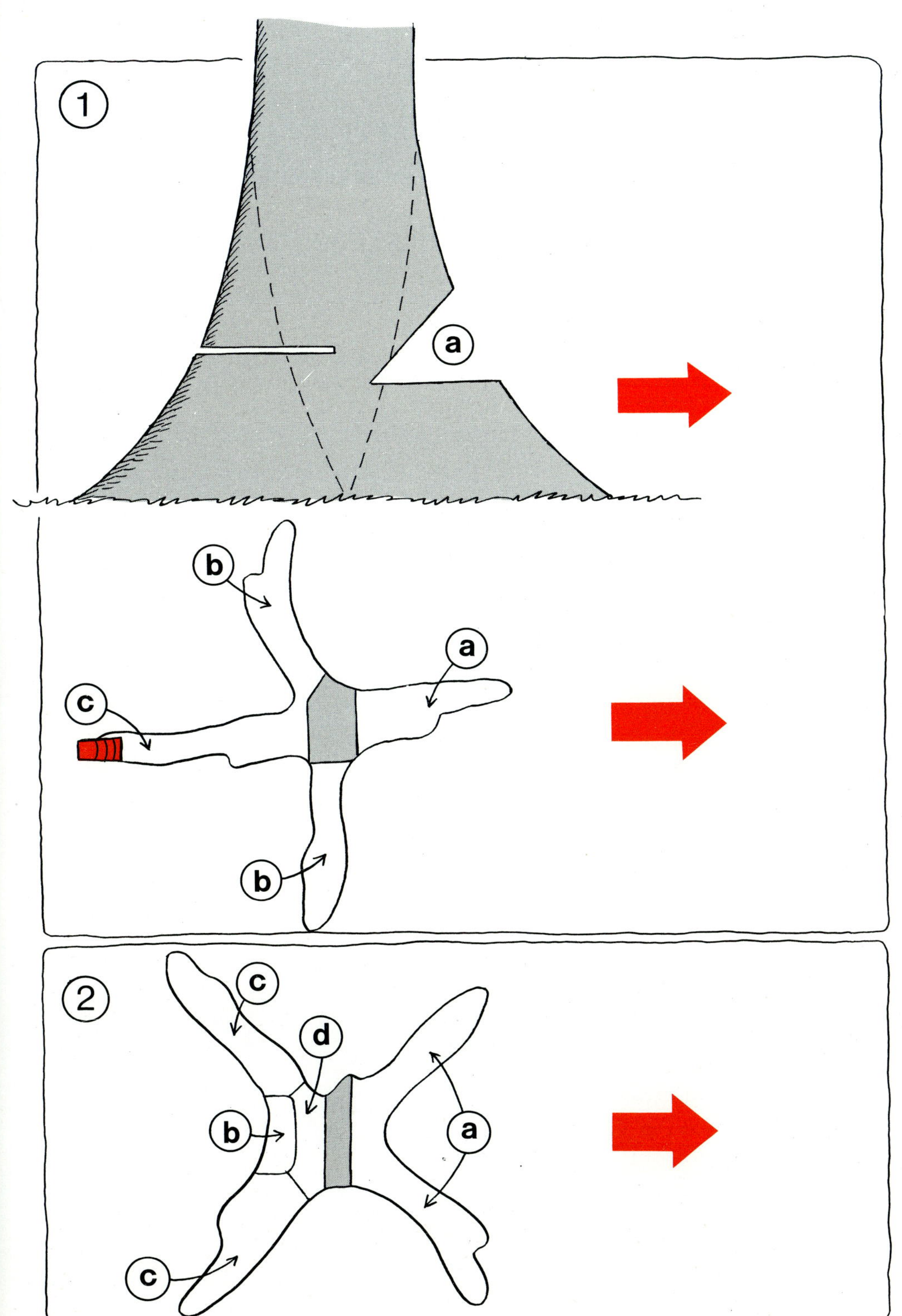
1
a
b
a
c
b
2
c
d
b
a
c

7.8 LEANING TREES WITH PLANK BUTTRESSES

1. Trees leaning into the direction of fall

   Such trees are dangerous to fell because they start to fall much faster than vertical trees, leaving less time to retreat. Furthermore, the wood is under much tension and may split if not cut properly.

   First the undercut (a) is made. The saw should not penetrate too deeply into the tree to avoid saw pinching. The backcut (b) starts with a boring cut behind the hinge and proceeds to the outside of the buttresses leaving some wood to act as an anchor (c) (d) and (e).

   This anchor wood is finally sawn by oblique cuts from the outside, in the order (c), (d), (e).

2. Trees leaning to one side

   Such trees can be directed to fall at an angle on either side of the lean. The method is similar to that for cylindrical trees.

   The undercut forms a right angle with the felling direction (a).

   When making the backcut (d) (e), a narrower hinge is left on the side of the lean (b).

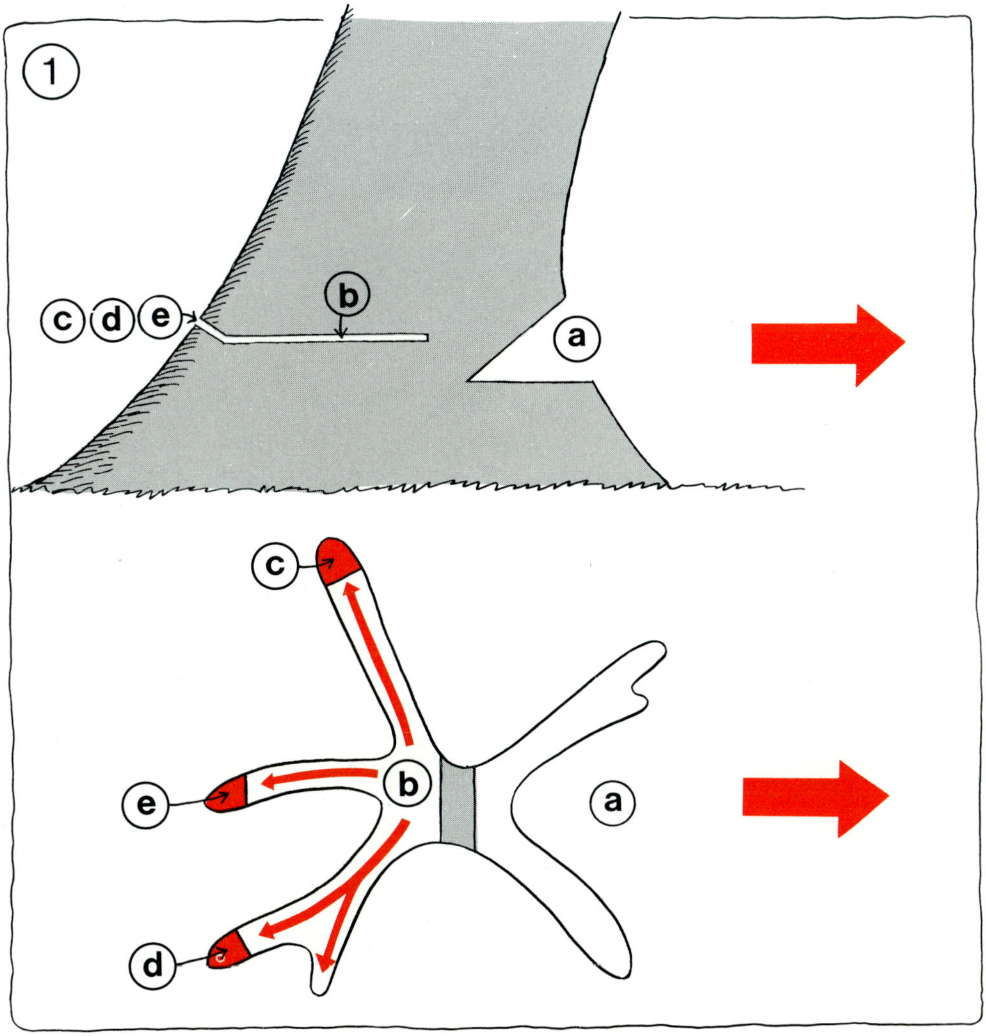

1

c d e

b

a

c

e

b

a

d

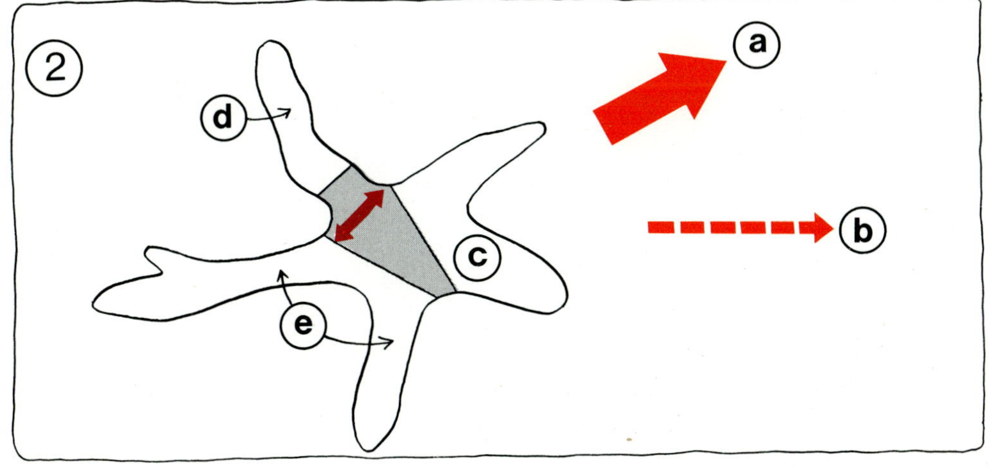

8. RELEASING LODGED TREES

8.1 USE OF SIMPLE TOOLS

When thinning man-made forests falling trees are frequently stopped by other trees. The tree is then said to be lodged and is called a hangup.

Skilled operators will try to avoid this by felling the tree to open spaces. A proper undercut, an adequate hinge and wedging will help to reduce lodging, but this will not always be avoided.

Beware - dislodging hangups can be very dangerous. Think first before deciding how to take the tree down. Do not walk or work below a hangup. Do not try to fell the tree which is holding the lodged tree. Do not fell another tree on to the hangup. Do not climb the lodged tree to loosen its crown.

Recommended techniques for small trees

Place suitable material (branches, poles) on the ground on to which the tree might slide backwards.

1. Cut the remaining wood which may still connect the tree with the stump, preferably by means of an axe. If the chainsaw is used it might easily get pinched.

2. Use the cant hook to roll the tree to one side.

3. Use a pole to push the butt end backwards.

4. Use a manual winch to pull the tree backwards.

1
2
3
4

8.2 USE OF SPECIAL EQUIPMENT AND MACHINES

1. If lodging (or hangups) occur very frequently, a sulky may be a convenient aid to lift the tree up from the stump and to pull it down. The sulky may also be used to skid it further to the skidding line. Sulkies can be made in local workshops. They are an excellent means of facilitating this heavy and dangerous job. Their use is restricted to trees up to a volume of about 0.5 cubic metres.

2. Heavy trees which cannot be dislodged by manual work should be pulled down with a skidding tractor. The tractor must be placed at a safe distance from the lodged tree and the cable winch used. This method is usually applied in natural tropical forests where lodging seldom occurs.

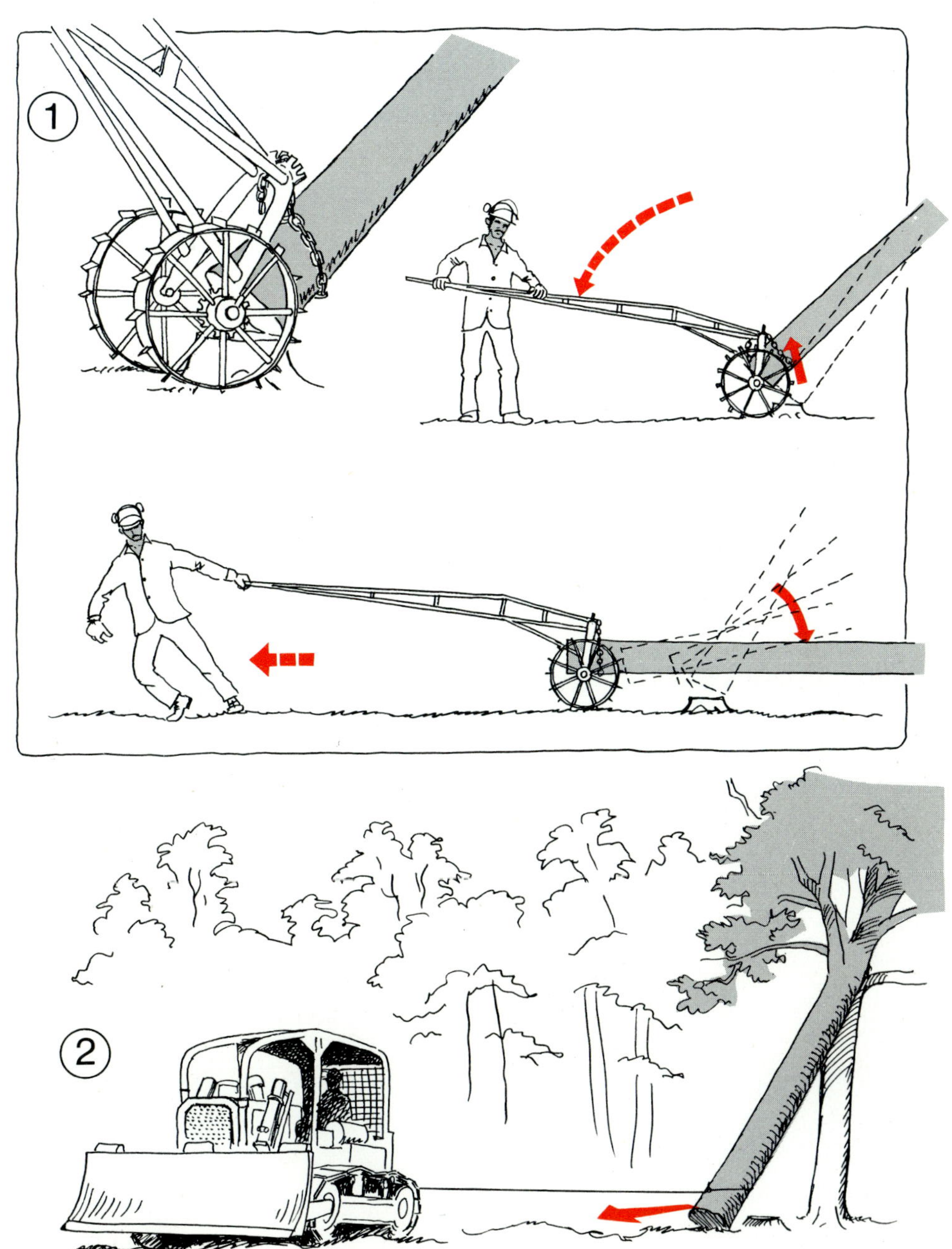
1
2

9. DELIMBING

9.1 BASIC RULES

1. Stand in a safe working position and watch out for obstacles.

2. Keep your eye on the chainsaw and if possible support its weight with your thigh.

3. Adjust the grip on the handle to the position of the chainsaw.

4. If possible, let the tree support the weight of the chainsaw.

5. Use the chainsaw as a lever by using the spikes as an anchor.

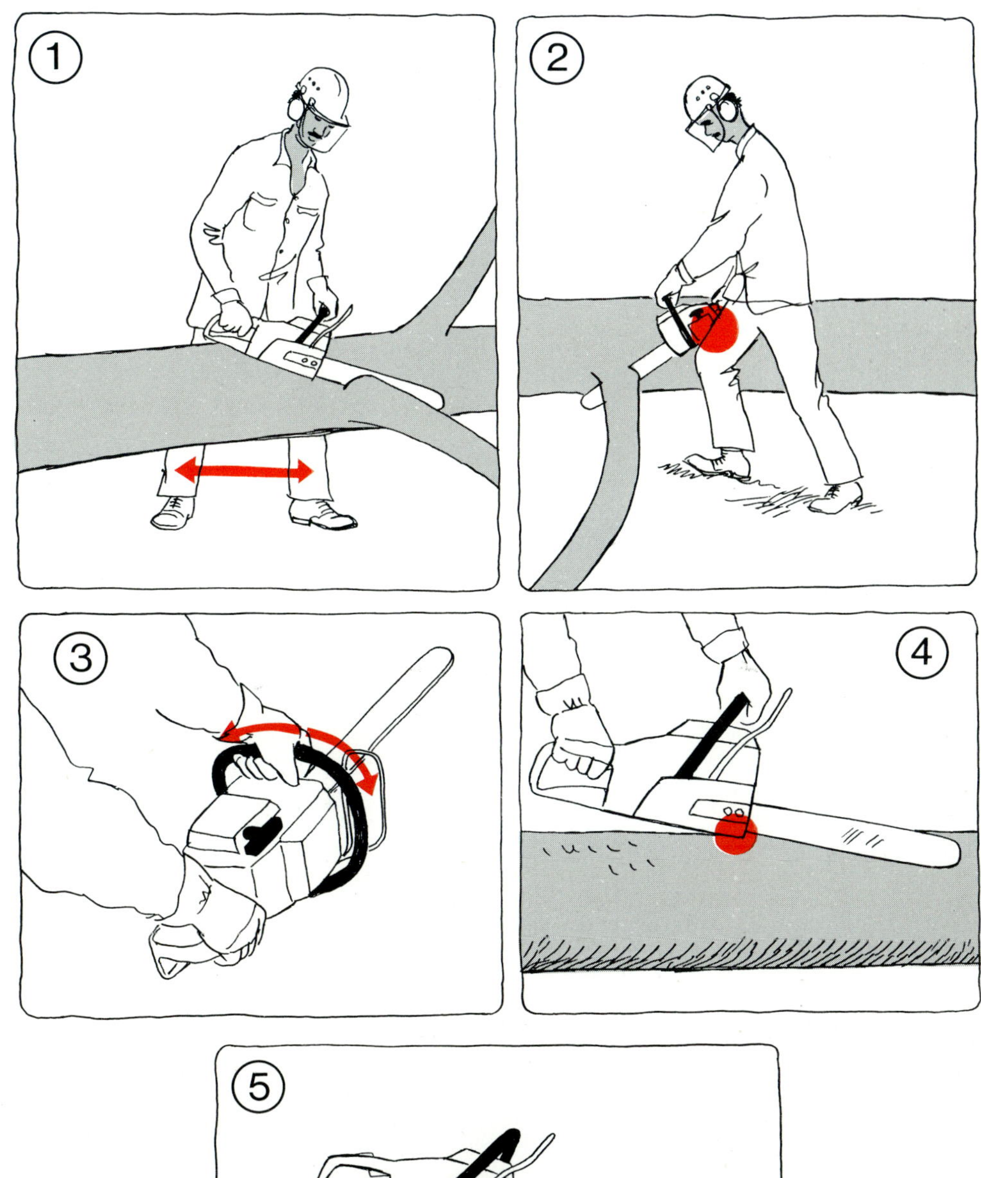
1
2
3
4
5

9.2 DELIMBING SMALL OR MEDIUM SIZED CONIFERS

Where the wage level is low it will be preferable to use an axe for delimbing.

Delimbing with a power saw requires highly skilled operators.

It is recommended that a regular work pattern, following the rings of branches, be adopted.

The operator stands on the left side of the tree. He works from the butt to the top of the tree. The saw moves from the right to the left on the first ring (1) (2) (3), and then moves to the next ring, this time cutting from left to right (4) (5) (6). This technique requires that the operator does part of the delimbing with a forward-running chain (1) (2) (4) (5), and the remainder with a backward-running chain (3) (6).

When the tree lies in a hollow, the branches on the underside of the two rings are cut in one movement before the operator moves forward to the next two rings (7). When the tree lies on the ground it is turned when the operator has worked his way to the top. The remaining branches are cut while the operator returns to the butt of the tree (8).

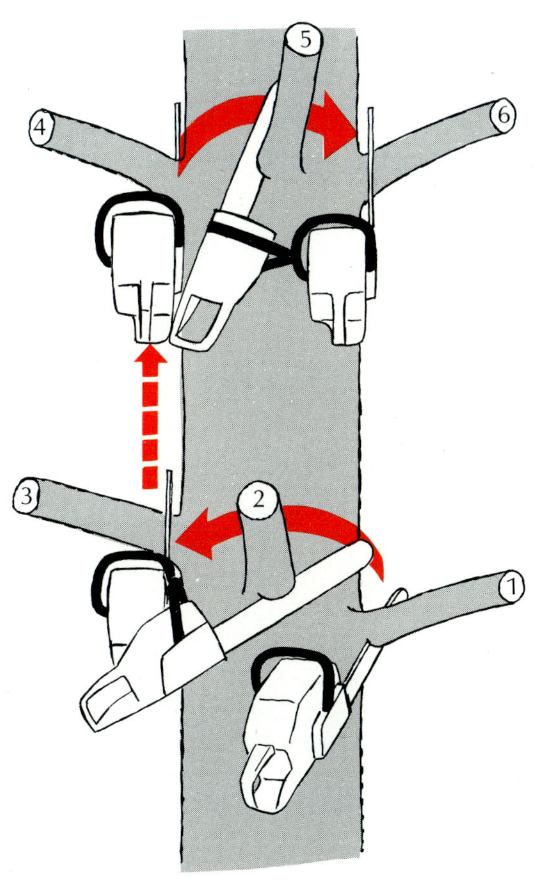
5
4
6
3
2
1

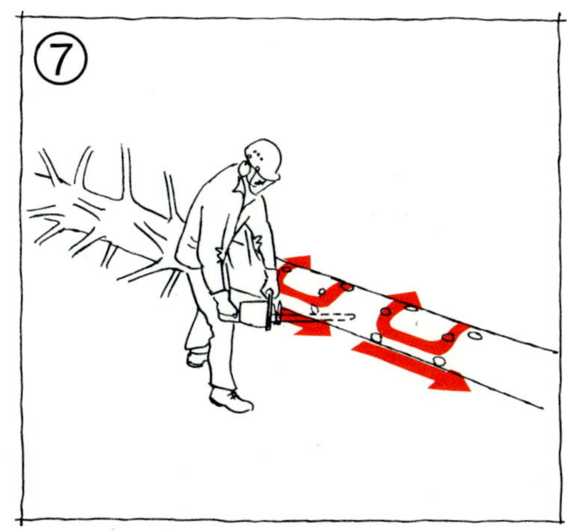
7

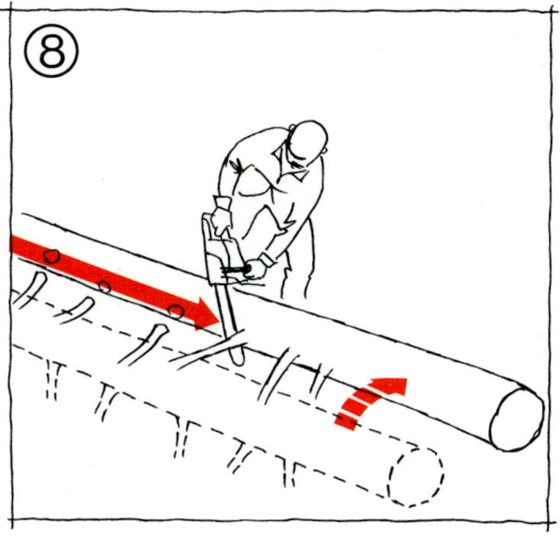
8

9.3 DELIMBING OF LARGER TREES

This operation requires much attention to avoid:

- the guide bar becoming pinched
- the wood cracking
- the operator being injured by branches swinging back or dropping down, or by the tree shifting.

The following rules should be observed:

1. - first cut and remove branches hindering your work.
   - cut branches in two or more sections when there is a danger of cracking at the base or when this facilitates clearing work (a) (b) (c).
   - keep your working space clear of cut branches.

2. It is important to observe tension in the wood. On larger branches, first cut the side which is under compression (a). Withdraw the guide bar before it gets pinched. Then cut from the other side (b).

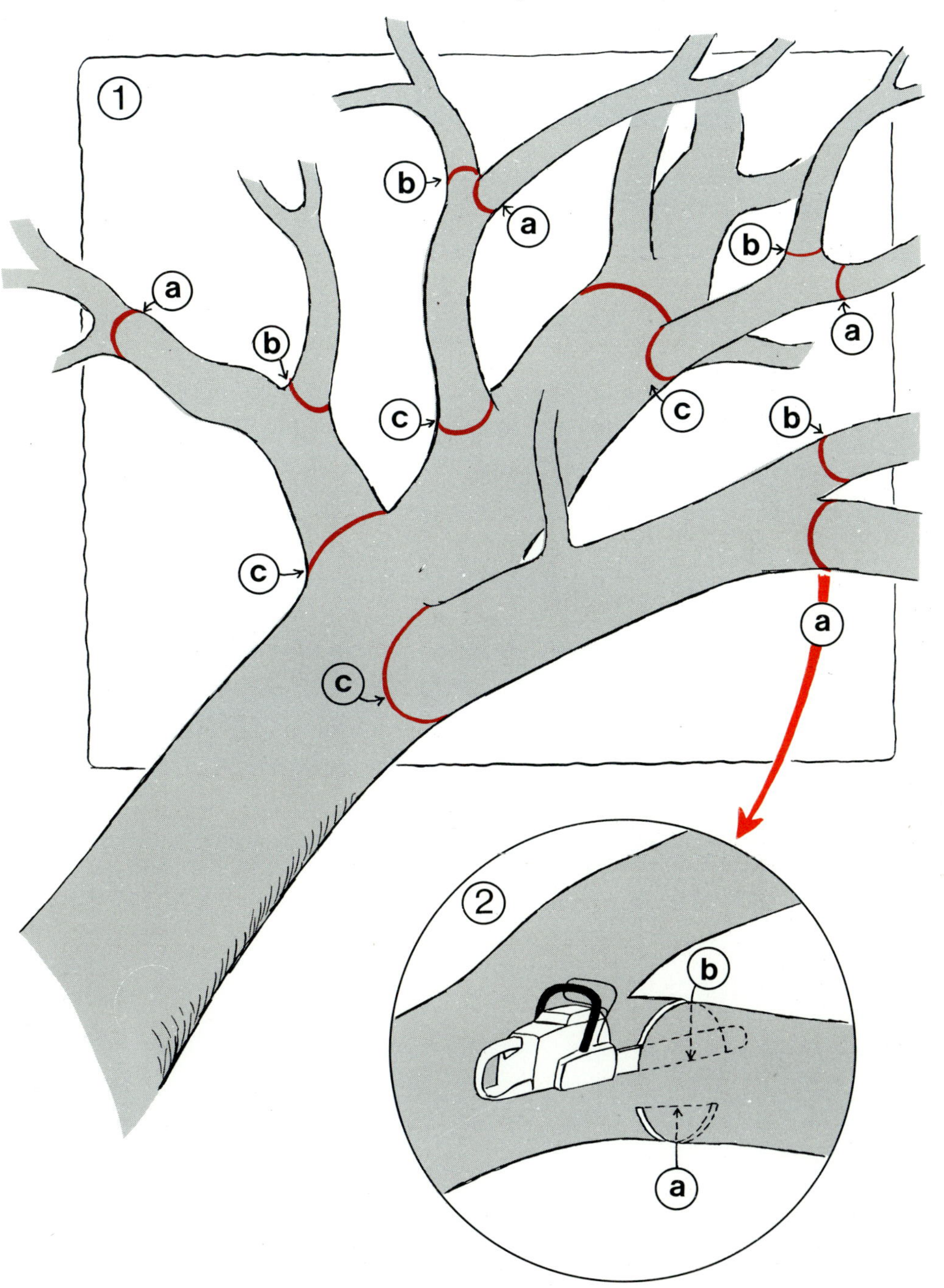
1
a
b
c
b
a
b
a
c
c
b
a
c
a
2
b
a

10. CROSS-CUTTING

10.1 BASIC RULES

Felling, delimbing and cross-cutting should be done by the same team in one continuous operation, tree after tree.

During cross-cutting, as in delimbing larger trees, the operator must observe whether or not the guide bar will become pinched in the cut and whether the log will move towards him when the cut is completed.

1. Always stand on the safe side, especially in hilly terrain.

2. For small trees the operator does not need assistance in cross-cutting. Small trees are cross-cut in one continuous movement from one side (a). The setting of one wedge is usually sufficient to avoid pinching of the guide bar (b).

3. For big trees the assistant marks the tree to be crosscut together with the operator while the helper clears the area around the cutting place. The assistant watches the saw cut, drives wedges and takes over the saw when the operator becomes tired.
   When the tree is too big for the guide bar, cross-cutting is done from two sides. The saw is shifted several times (a) (b) (c) (d). In this way trees with diameters up to twice the length of the guide bar can be cross-cut.
   It is important to have wedges available in case the guide bar gets pinched. For big trees, two wedges should be used to avoid the tree swinging to one side (and therefore pinching the guide bar) before the cut is finished (e). If there is a risk of pinching, wedges should be inserted as soon as the cut is deep enough.

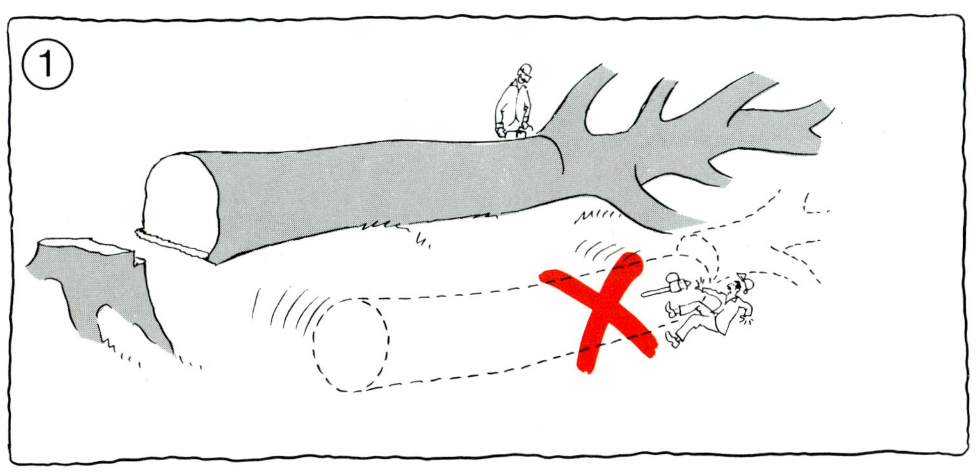
1

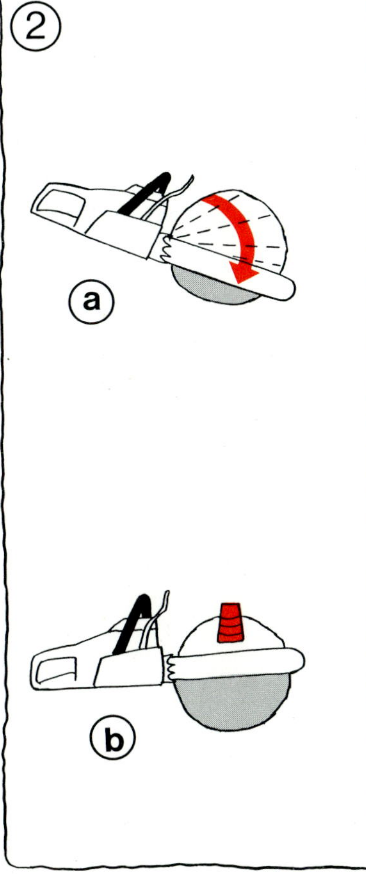
2
a
b

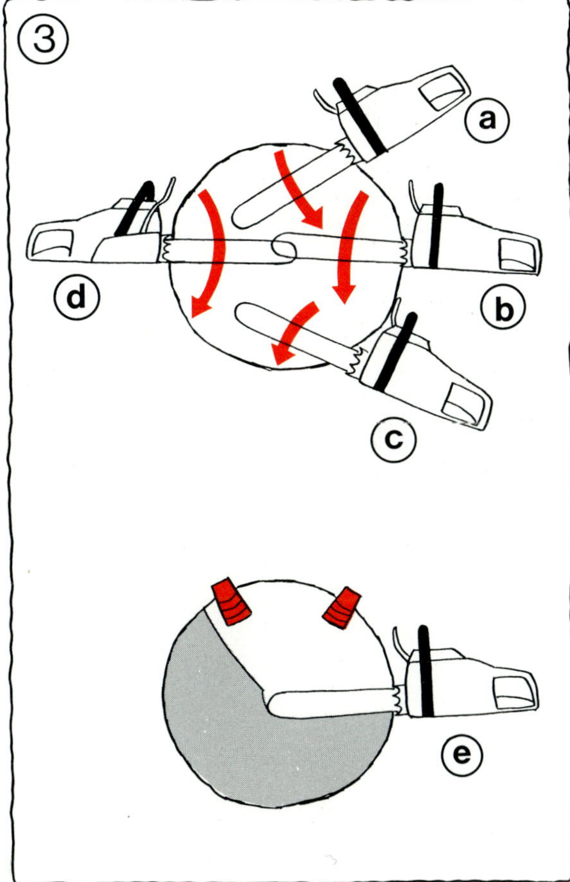
3
a
d
b
c
e

10.2 CROSS-CUTTING TREES UNDER TENSION

1. If a small tree is under tension, first make a shallow cut on the side which is under compression (a). Withdraw the saw before it gets pinched. Then cut from the tension side (b).

2 & 3. For larger trees under tension, first make cut (a). Next cut on the side which is under compression (b). Carefully observe the opening (closing) of this cut and withdraw the bar before the cut closes and pinches the bar and chain.
Finally, make cuts (c) and (d).
In this manner, wood losses through splitting, and pinching of the saw can be avoided.

There are variations of this technique, such as when the tree touches the ground. In this case it may be necessary to apply a boring cut. But the principle is always the same:
First cut slightly on the side which is under compression.
Leave some holding wood.
Then cut the side which is under tension.

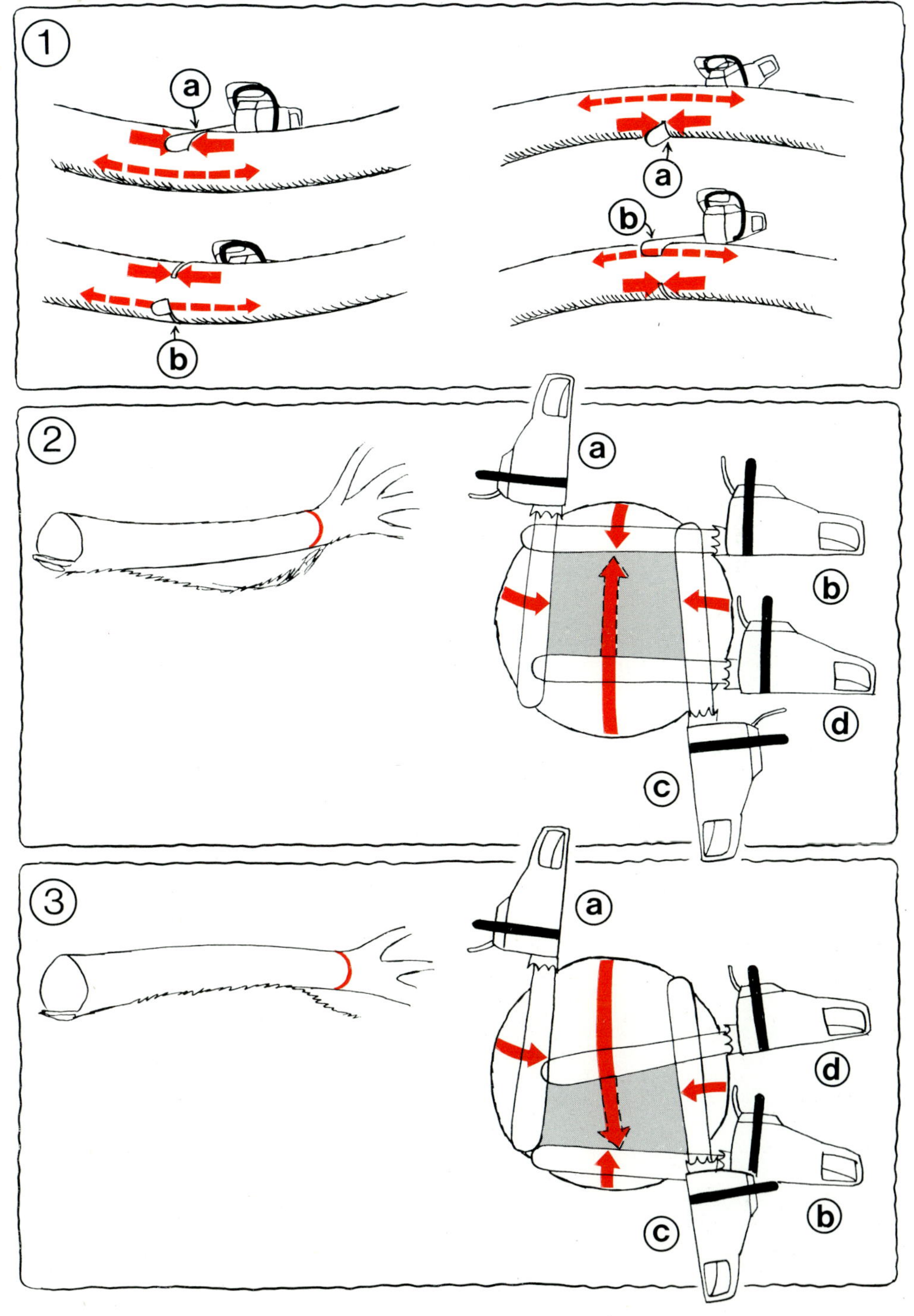
1
a
a
b
b
2
a
b
d
c
3
a
d
b
c

10.3 CROSS-CUTTING WINDFALLS

Windfalls are dangerous to cross-cut. The wood is often under high tension. The working place is sometimes narrow and the tree difficult to get at. Stumps may tip forward or backward when cut loose from the trunk. Therefore the operator must be well trained and experienced. Other people should not be near him when he is cutting.

In dense wind-felled stands the trees are first cut loose from their stumps. Further work is then made much easier if they are then pulled into open places for delimbing and cross-cutting.

1. Dangerous stumps must be prevented from overturning by use of supports (a) (b), or by cable (c) before sawing starts.

2 & 3. During cutting, care must be taken that the tree can easily separate from the stump.

The first cut is made on the compression side (a), followed by the second cut from above (b).

The second cut should be 2 to 5 cm nearer to the stump if the tree is expected to swing upwards (2).

If the tree is expected to drop downwards the second cut should be 2 to 5 cm nearer the top (3).

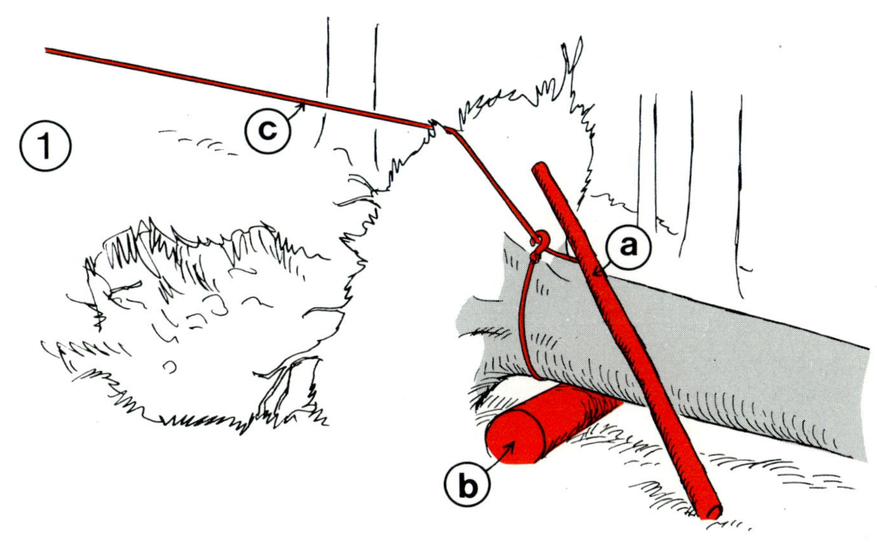
1
c
a
b

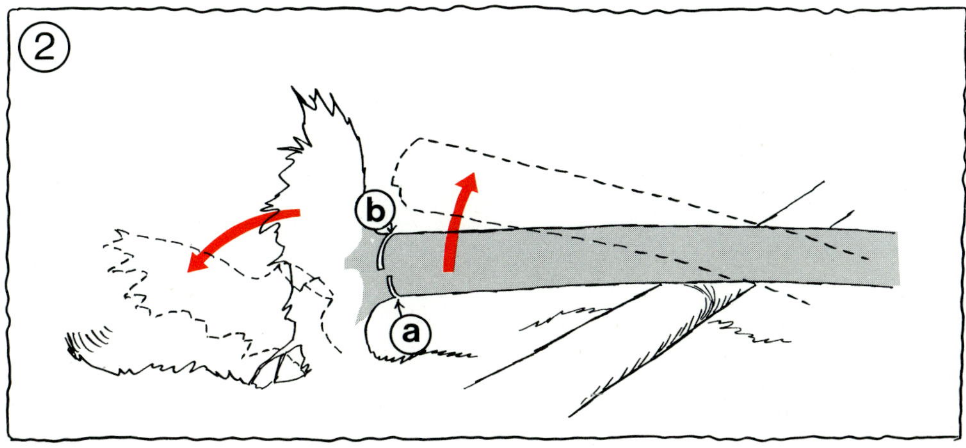
2
b
a

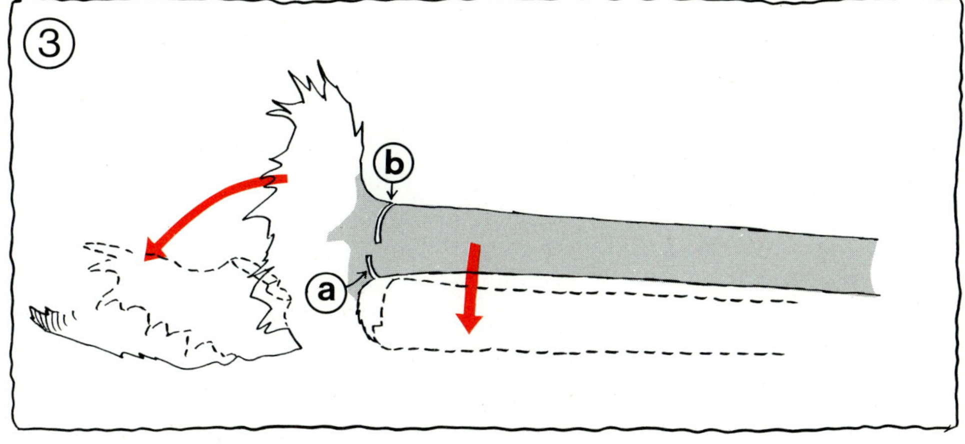
3
b
a

11. WOOD WASTAGE DUE TO POOR WORKING TECHNIQUES

11.1 WOOD WASTAGE IN FELLING

1. High stumps are an indication of poor workmanship and insufficient supervision. Sometimes they are the result of putting felling marks, which are to be left on the stump for control purposes, too high. Sometimes workers find it more comfortable to cut about 1 m above ground level.
   Except in special cases (e.g. hollow or heavily buttressed trees) the stump should be as low as possible. This helps to avoid wastage of wood and because of the lower stumps makes skidding easier.

   Where the wage level is low and the timber price high the value of wood left in one stump may correspond to a week's wages of the operator, or more.

2. Considerable wood losses in felling can also occur if the tree is felled without an undercut or with an insufficient undercut. If the undercut is at the same level as, or higher than the backcut there is a risk that wood fibre will be pulled out of the butt end, reducing the value of the log.

3. If the undercut is too small, this can be most dangerous because the tree's fall is no longer guided properly.

   If the undercut is not made properly the tree can split.

   At the same time considerable wastage occurs on the valuable butt end.

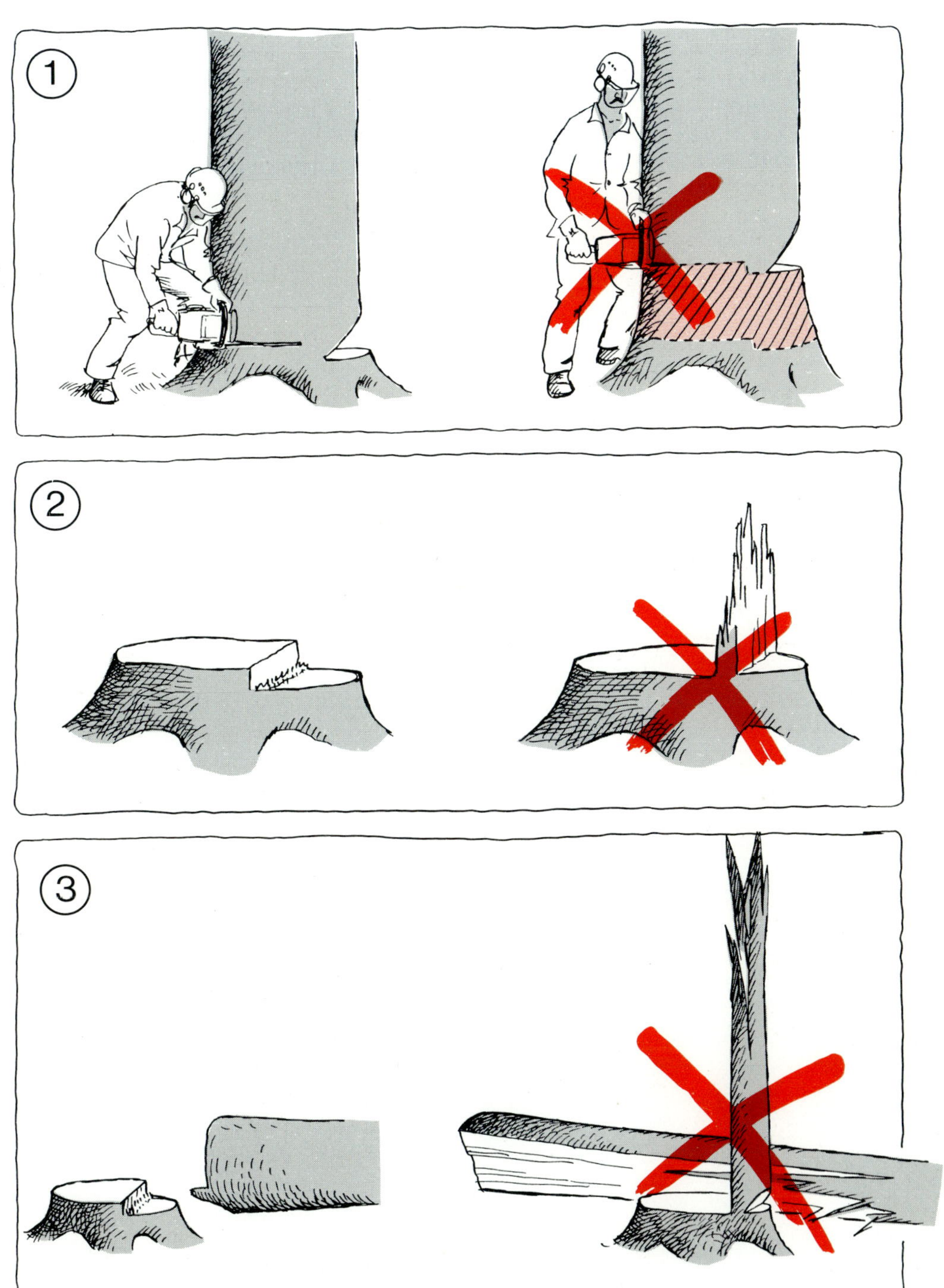
1
2
3

11.1 Wood Wastage in Felling (Continued)

A considerable amount of valuable wood is lost when large trees are felled across obstacles on the ground such as hollows (1), ridges (2), logs (3) or rocks. Most species break if they hit such obstacles. Although the broken part may be small, the degrade caused by cutting out the break can be considerable.

The experienced operator will carefully look for obstacles. When determining the felling direction he will try to avoid them. This will be possible in many cases. Remember that even heavily leaning trees can be felled to a point about 30 degrees on either side of the lean.

Efforts to avoid obstructions not only reduce waste but also facilitate work because unnecessary cross-cutting is also avoided.

For valuable hardwood species the breaking up of a tree can be very costly (wasteful).

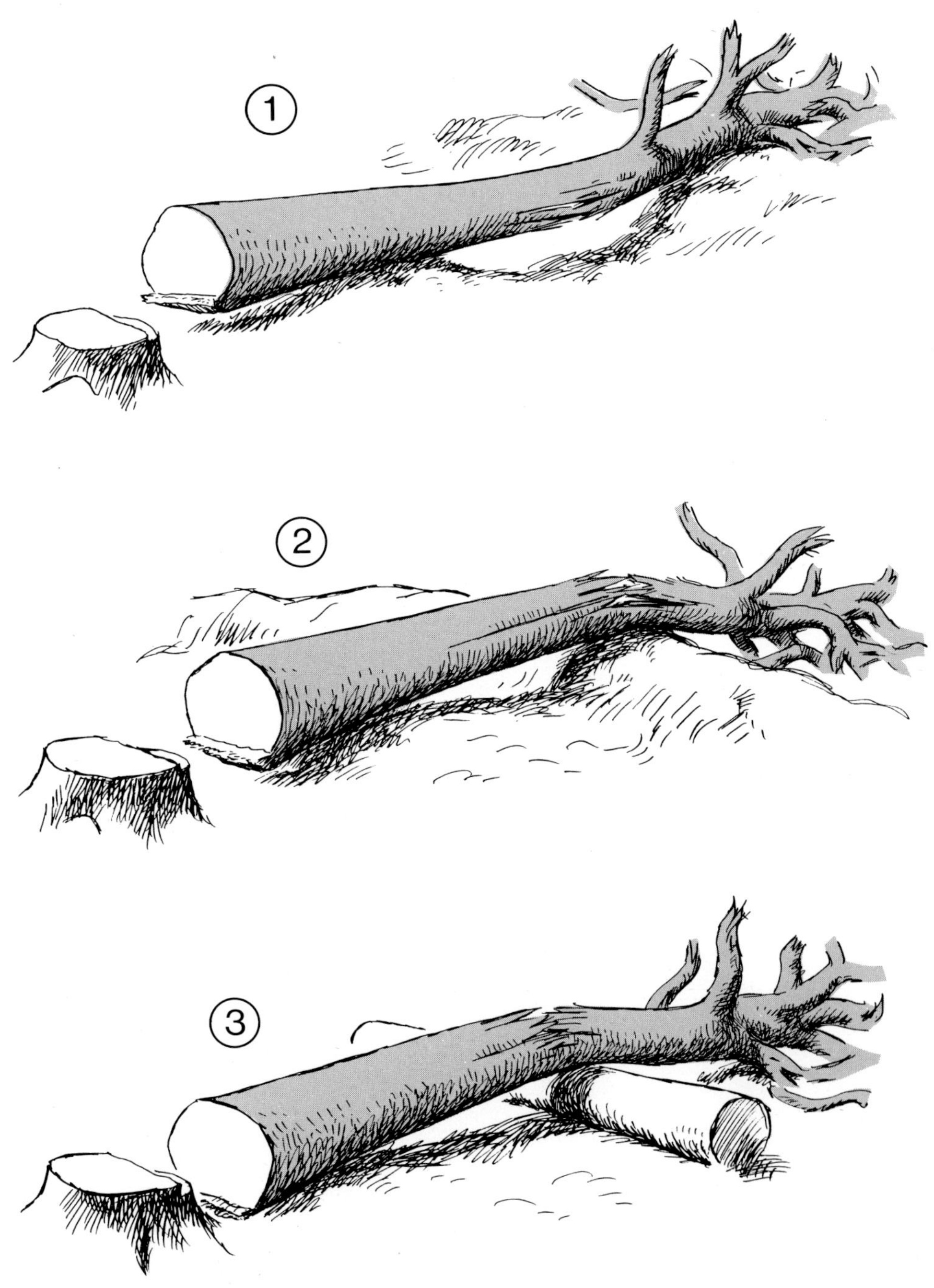
1
2
3

11.2 WOOD WASTAGE IN CROSS-CUTTING

1 & 2. When trees are under tension they will easily split if cross-cutting starts on the side which is under tension.

Proper sawing techniques with the chainsaw reduces this danger to a large extent.

Always remember to cut first on the side of the tree which is under compression (a). Then complete the cut on the side which is under tension (b).

3. When cutting windfalls from the stump, splitting of wood can easily occur. Usually the upper side is under tension. If the cut is started here, in many cases the butt log will split up to several meters (a).

The underside, which is under pressure, must be cut first. If this side is not accessible because the tree is laying on the ground, a spade should be used to make a hole underneath (b).

A boring cut can also be made to cut the underside but care must be taken that the chain does not touch the ground.

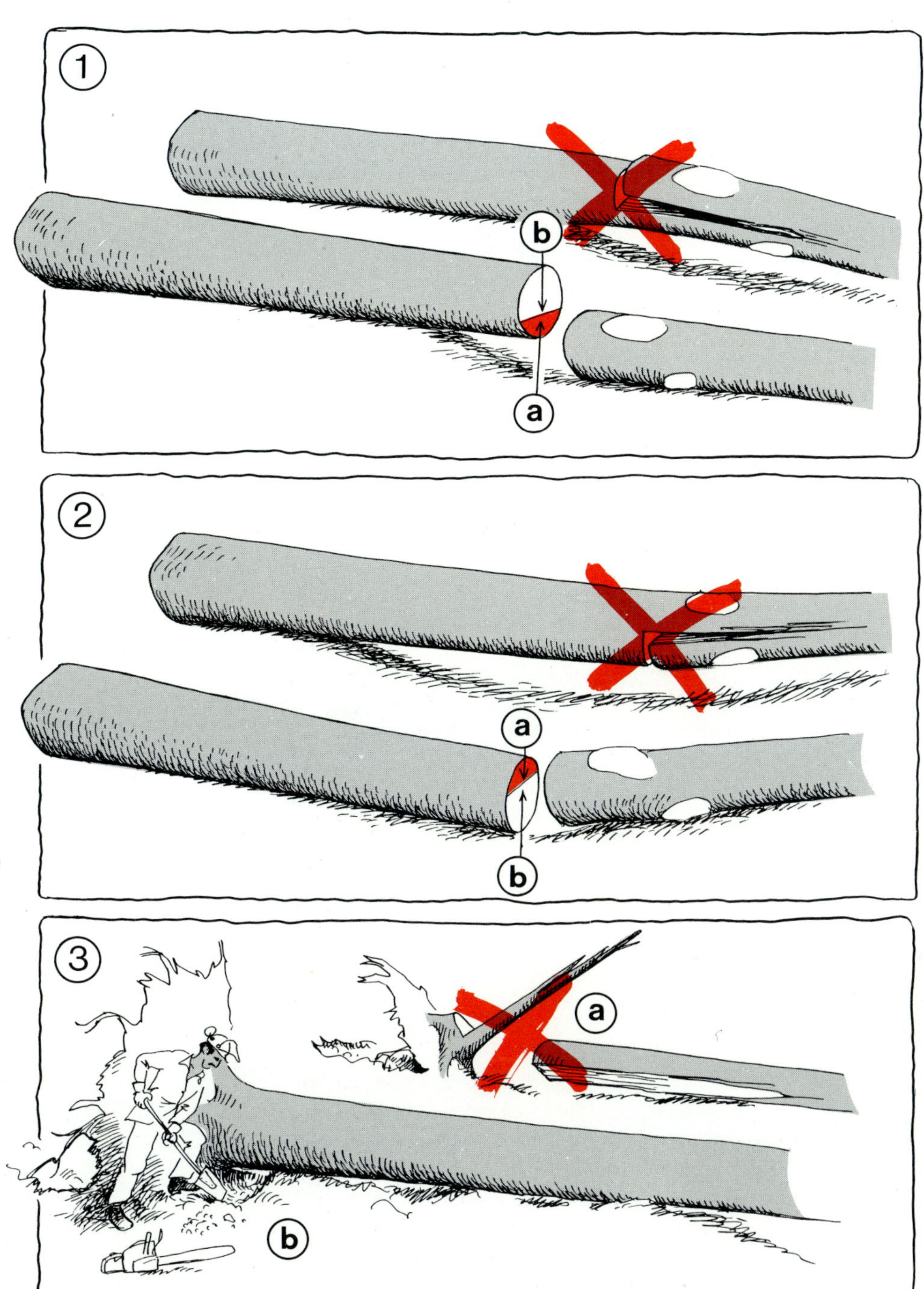
1
b
a
2
a
b
3
a
b

12. MAINTENANCE OF SAW CHAIN AND GUIDE BAR

12.1 THE SAW CHAIN

The chain must always be sharp and well maintained.

A blunt and poorly maintained chain requires more time, effort and fuel to do the job. It wears out very rapidly, will damage the guide bar, and may even overwork the engine. It may also increase the danger of accidents, because kickback may increase; faster cutting reduces exposure time to dangerous situations; and increased fatigue makes the worker more accident prone.

The operator must know when the chain is not in good order. He must be able to sharpen it in the forest.

The different parts of the chain are:

1. Cutter link, consisting of:
   - top plate (a)
   - cutting edge (b)
   - side plate (c)
   - depth gauge (d).

2. Drive link

3. Tie strap

4. Rivet

5. Safety links. There exist different types of safety links which reduce the danger of kickback. Safety chains with such links are especially recommended for cutting and delimbing small trees.

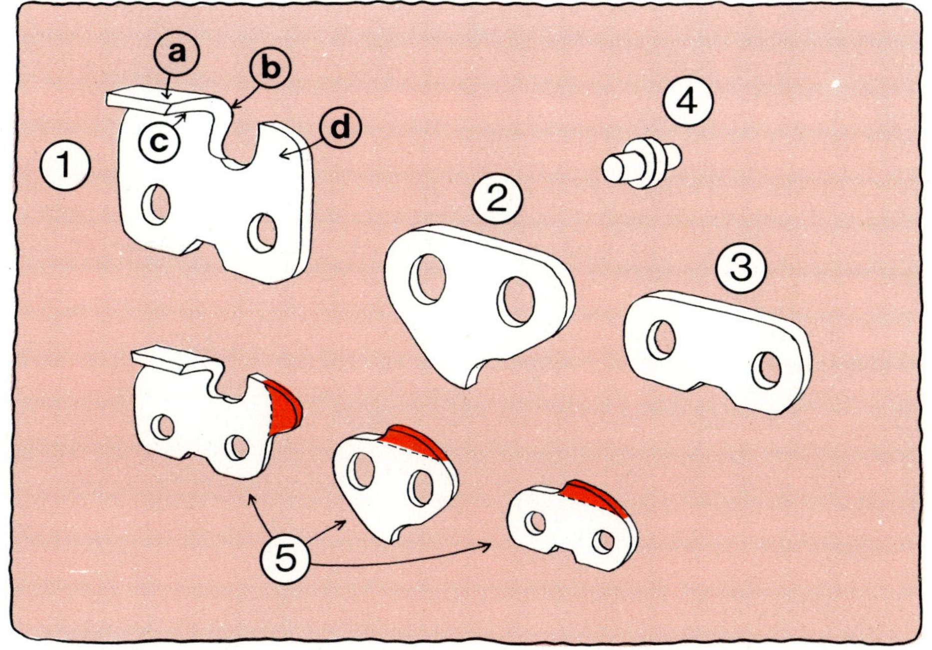
a
b
c
d
1
2
3
4
5

12.2 SHAPE OF CUTTER

The most common types of cutters have a:

round side plate (1), or a

chisel shaped side plate (2).

The correct top plate filing angle (3) is 35 degrees for the round side plate and 30 degrees for the chisel shaped side plate.

The correct side plate filing angle (4) is 90 degrees for the round side plate and 85 degrees for the chisel shaped side plate.

To maintain the correct filing angles, it is necessary that a round file be used. The diameter of this file must correspond to the size of the chain. The size of the chain is called the pitch. The pitch is the distance over three rivets, divided by two (5).

The most common pitch and the corresponding diameter of the round file for a new chain are as follows:

Pitch		Diameter of round file		Size of saw
inch	mm	inch	mm	
1/4	6.35	1/8	3.8	light
0.325	8.26	3/16	4.8	light
3/8	9.23	7/32	5.5	medium
0.404	10.26	7/32	5.5	medium
1/2	12.70	1/4	6.4	heavy

If the cutter is filed back half its length a round file with a diameter of about 0.5 mm less than for the new cutter should be used. As a general rule 1/10th of the diameter of the round file should be above the top plate (6).

The depth gauge controls the depth of the cut. The depth gauge setting (7) depends on the pitch of the chain and the hardness of the wood. It usually varies from 0.025" (0.63 mm) to 0.030" (0.75 mm), to 0.040" (1.0 mm).

The more powerful the saw and the softer the wood the deeper the depth gauge can be set.

	①	②
③ 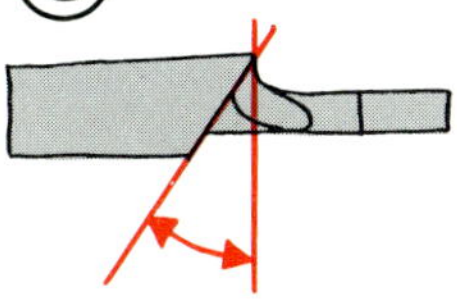	35°	30°
④ 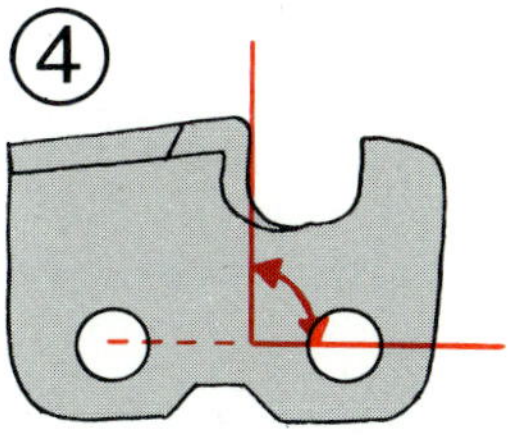	90°	85°

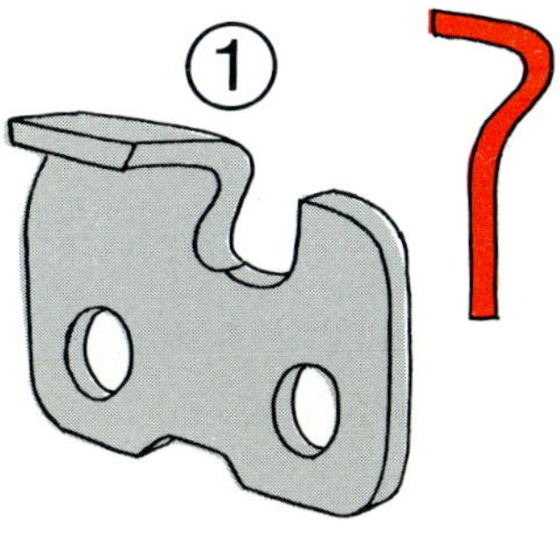

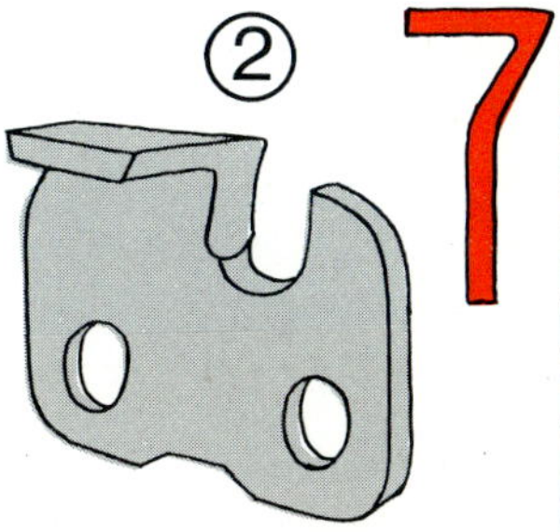

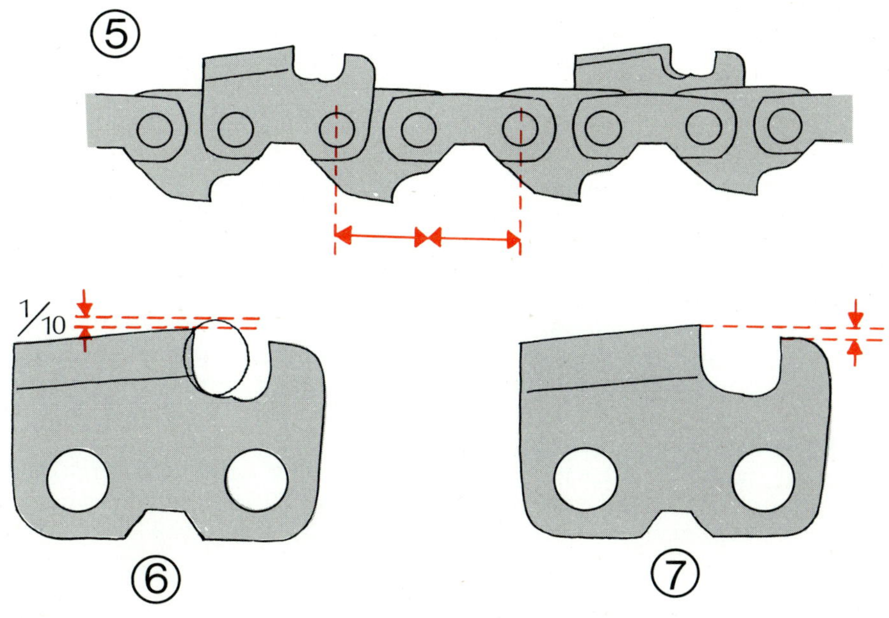

12.3 FILING EQUIPMENT

1. For filing in the forest a filing clamp (a), a round file (b) and a filing grid (c) with a magnet (d) will be sufficient.

   There are various types of file holders which may be used (e). These usually include a filing grid.

2. For comprehensive maintenance at the workshop a filing vice (a) with filing grid, a slide gauge (b), a depth gauge control (c) and a flat file for jointing (d) are needed.

3. If large numbers of saw chains are to be maintained in a central workshop an electric chain sharpener can be used. This machine must be so adjusted that all teeth get into the correct size and shape. Care must be taken with the machine that it does not sharpen more than necessary. Only light pressure should be applied, otherwise the chain will be used up too rapidly or might burn (overheat) and lose its hardness.

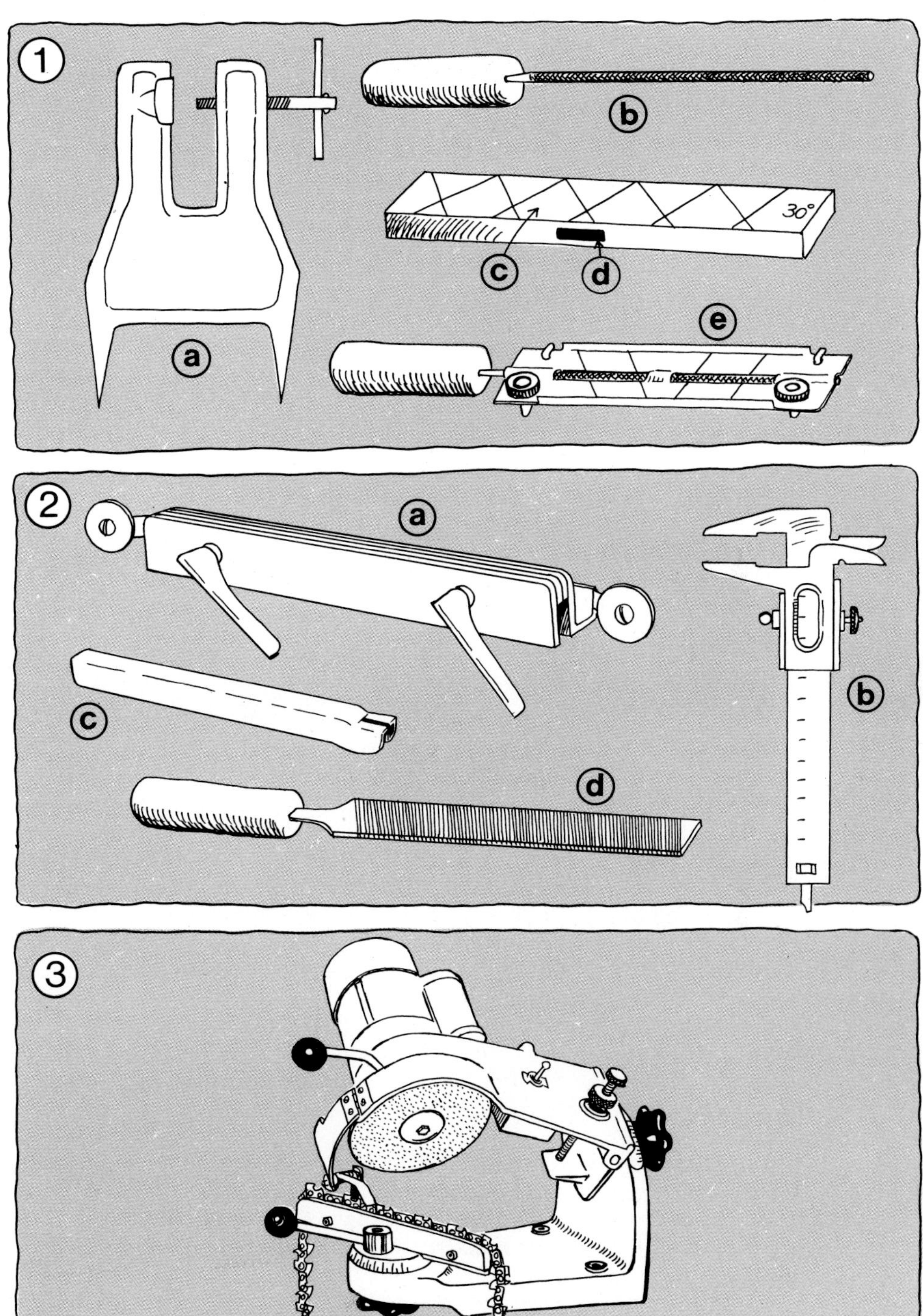
1
a
b
c
d
30°
e
2
a
b
c
d
3

12.4 CORRECT FILING OF CUTTER

1. When filing in the forest it is helpful if the saw can be fixed to a log or stump by means of a clamp (a) and also to attach a simple top plate filing grid to the guide bar (b) with a magnet.

2. The file, with the proper diameter, must move parallel to the top plate (a) and to the cutter (b) at the filing angle of 30/35 degrees (depending on the type of plate).

   After about one hour of sawing, two or three gentle strokes per cutter will keep the chain well sharpened.

   Always file in one direction, from inner to outer side of teeth (b). Never file from the outer to the inner side (c).

3. At least once a week, the chain must be carefully checked in the workshop. The same must be done when the saw chain has hit an obstacle and become blunt. Remove the chain from the guide bar and put it in a special filing vice or in an ordinary blacksmith's vice. First use a slide gauge (a) or improvise with a piece of wood or cardboard (b) to find the shortest tooth. Sharpen all teeth to the length of the shortest tooth. Make sure that the top plate angle and the side plate angle are correct.

4. When all cutters, on both sides, are sharp, check the depth gauge with a depth gauge control (a). Use a flat file to lower the depth gauge to the level of the control (b). Use the same file to round off front end of depth gauge (c).

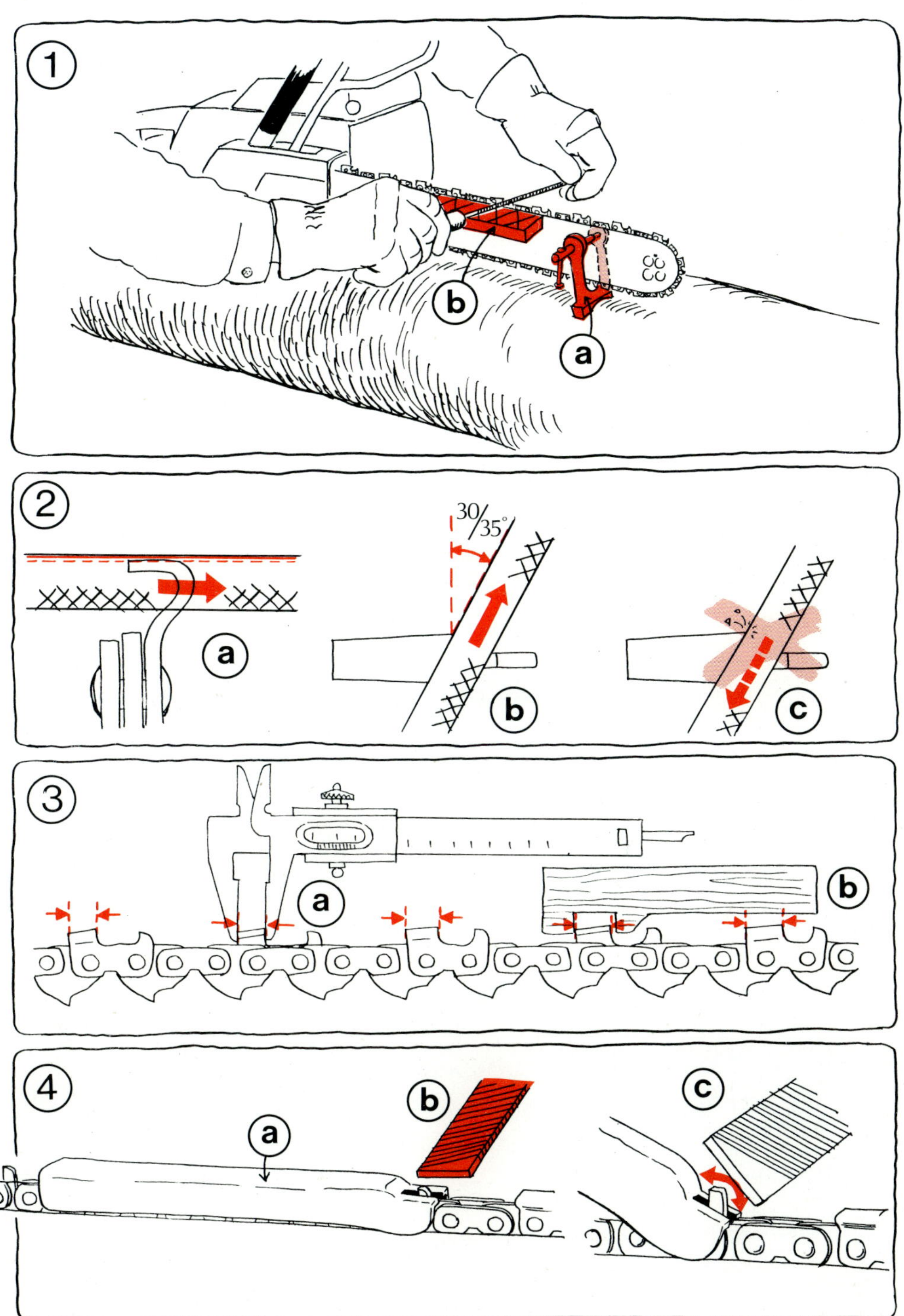
1
b
a
2
30/35°
a
b
c
3
a
b
4
b
a
c

12.5 RIVETING

If individual links of the chain become badly damaged or if the chain breaks, it will be necessary to replace the damaged link in the workshop.

1. Put the chain on the groove of either a riveting anvil (a) or a slightly opened blacksmith's vice. Do not place the cutter on the anvil (b).

   Knock out the rivets of the damaged link with a light hammer and punch (c).

2. File the new link to exactly the same size as the others in the chain.

3. Do not tap too heavily on new rivets because this will make the chain stiff.

   If available, a ballpean hammer should be used for riveting.

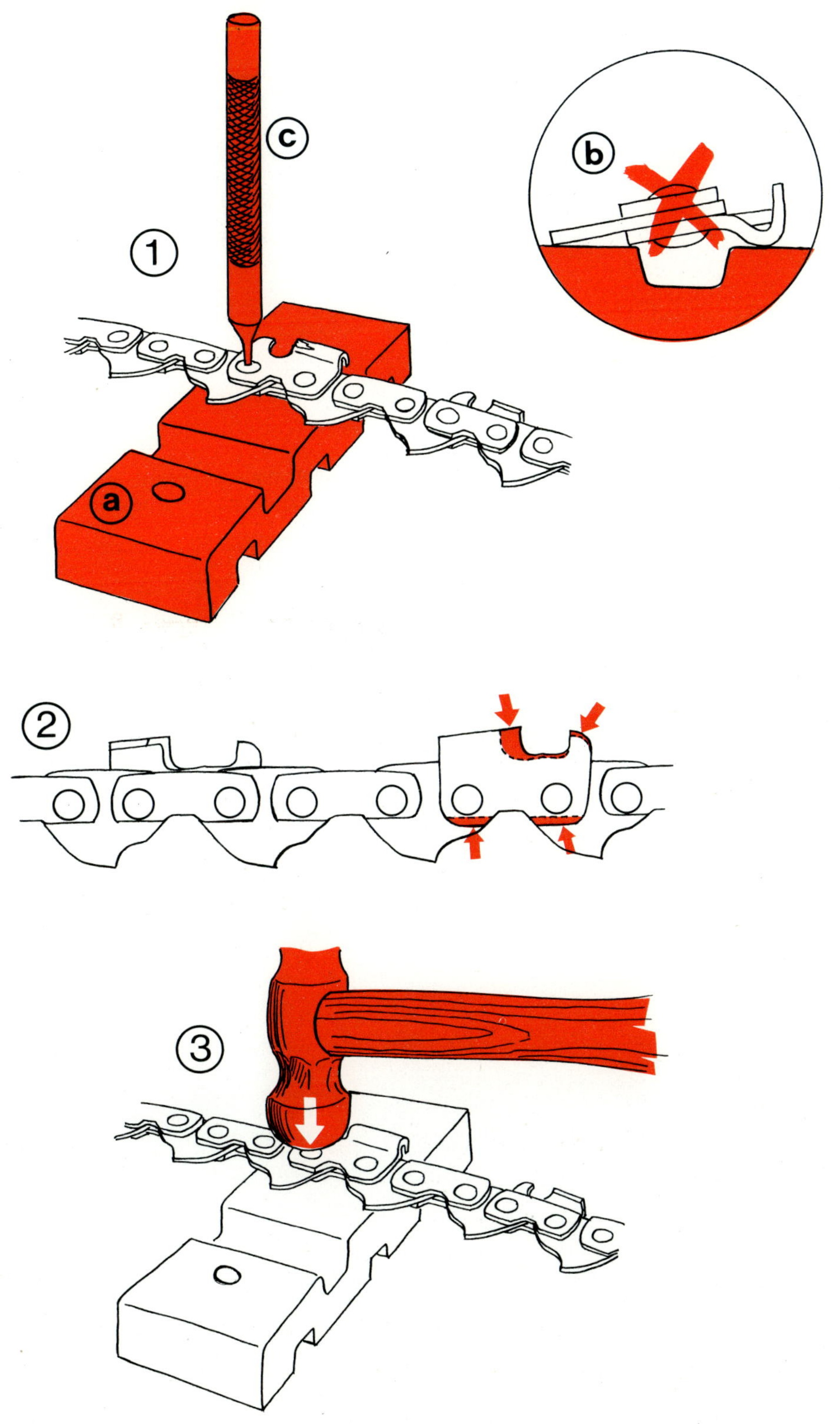
c
b
1
a
2
3

12.6 GUIDE BAR AND SPROCKET OR DRIVE RIM

1. To keep the saw chain in good condition the guide bar (a) and the sprocket (b), or drive rim (c), must also be in good order. The same applies to the sprocket nose (d) with which guide bars are often fitted.

2. Turn the guide bar at least once a day.

3. Clean the groove once a day.

4. Clean the chain oil holes once a day.

5. Clean the grease holes of the sprocket nose (a) and use a grease pump with bearing grease (b) until grease comes out of the sprocket.

6. At least once a week check the guide bar for burrs (a). Remove burrs with a flat file (b).

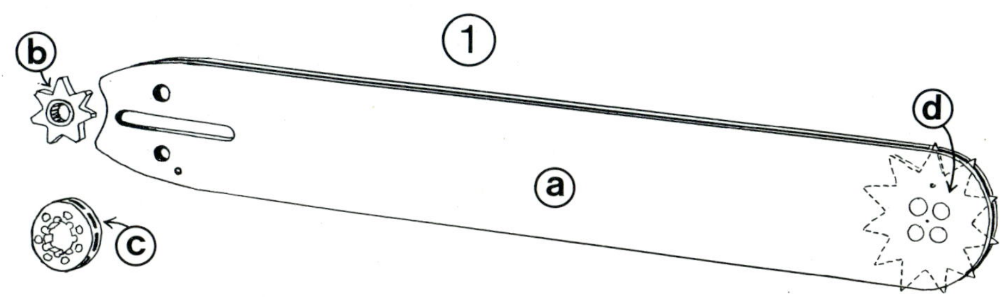
1
b
d
a
c

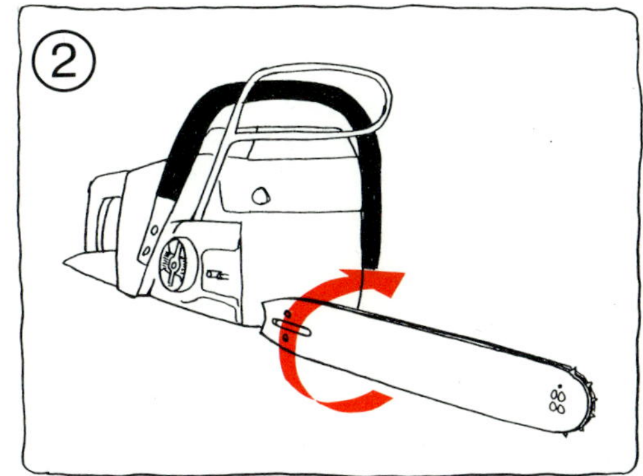
2

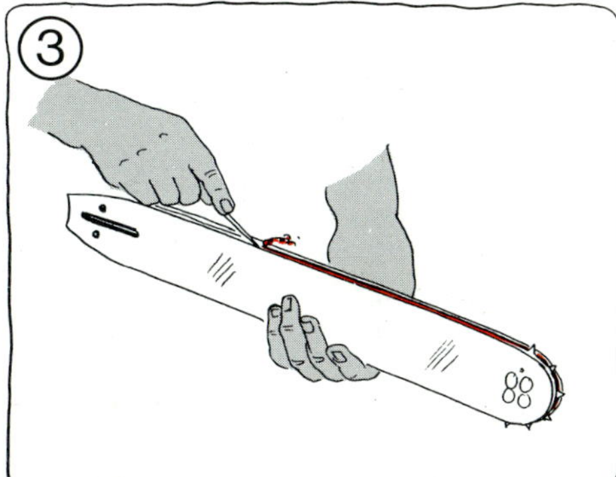
3

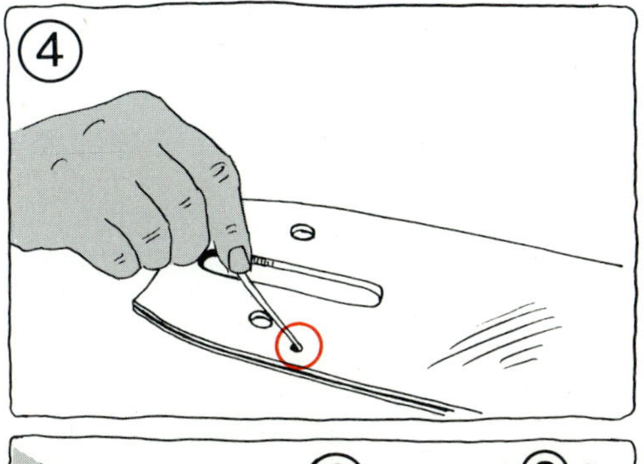
4

6
a
b

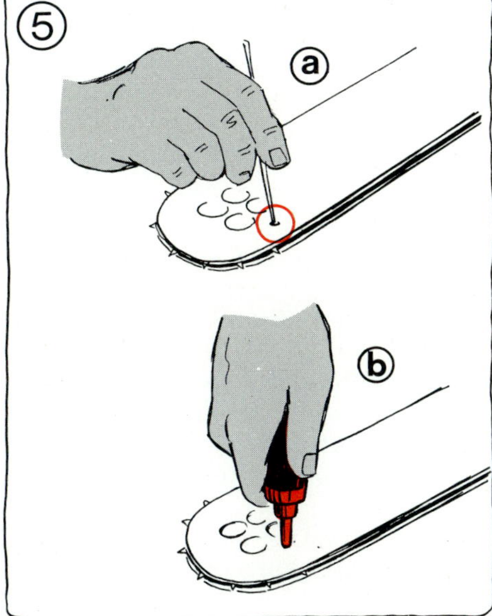
5
a
b

12.7 BASIC RULES FOR OPERATION OF A SAW CHAIN

1. Always make sure that the saw chain is under the correct tension. If the chain is too tight or too loose it will cause excessive wear.
   Lift the front end of the guide bar up when you tighten the guide bar bolts (a).
   Pull the chain towards the top of the guide bar (b).
   If the chain turns easily but remains tight, the tension is correct. If the chain is too loose it may jump off the bar, possibly damaging the chain, bar and sprocket.

2. Always check to see that the chain is receiving sufficient oil (a).
   If the chainsaw has provision for additional manual oiling, apply this when starting the saw or when cutting hardwood or large diameter trees (b). It may also be necessary to increase the automatic oiling.
   When refilling, always refill the oil tank (c) first; then refill the fuel tank (d).
   The size of the oil tank is such that it normally still holds some oil when the fuel tank is empty.

3. Before fitting a new chain on a saw, soak it in chain oil (a).
   Fit it on the guide bar, and adjust the tension (b).
   Then let it run slowly, without sawing, for about five minutes (c).
   Stop the engine and let it cool down (d).
   Adjust the tension (e).
   Let the chain run slowly again without cutting, as before (f).
   Repeat (d) (e) and (f) once or twice.
   Then make a few light cuts (g).
   Let the saw cool down again (h).
   Then adjust the tension on the chain (k).

   During the first hours of operation, frequently check the tension and oiling.

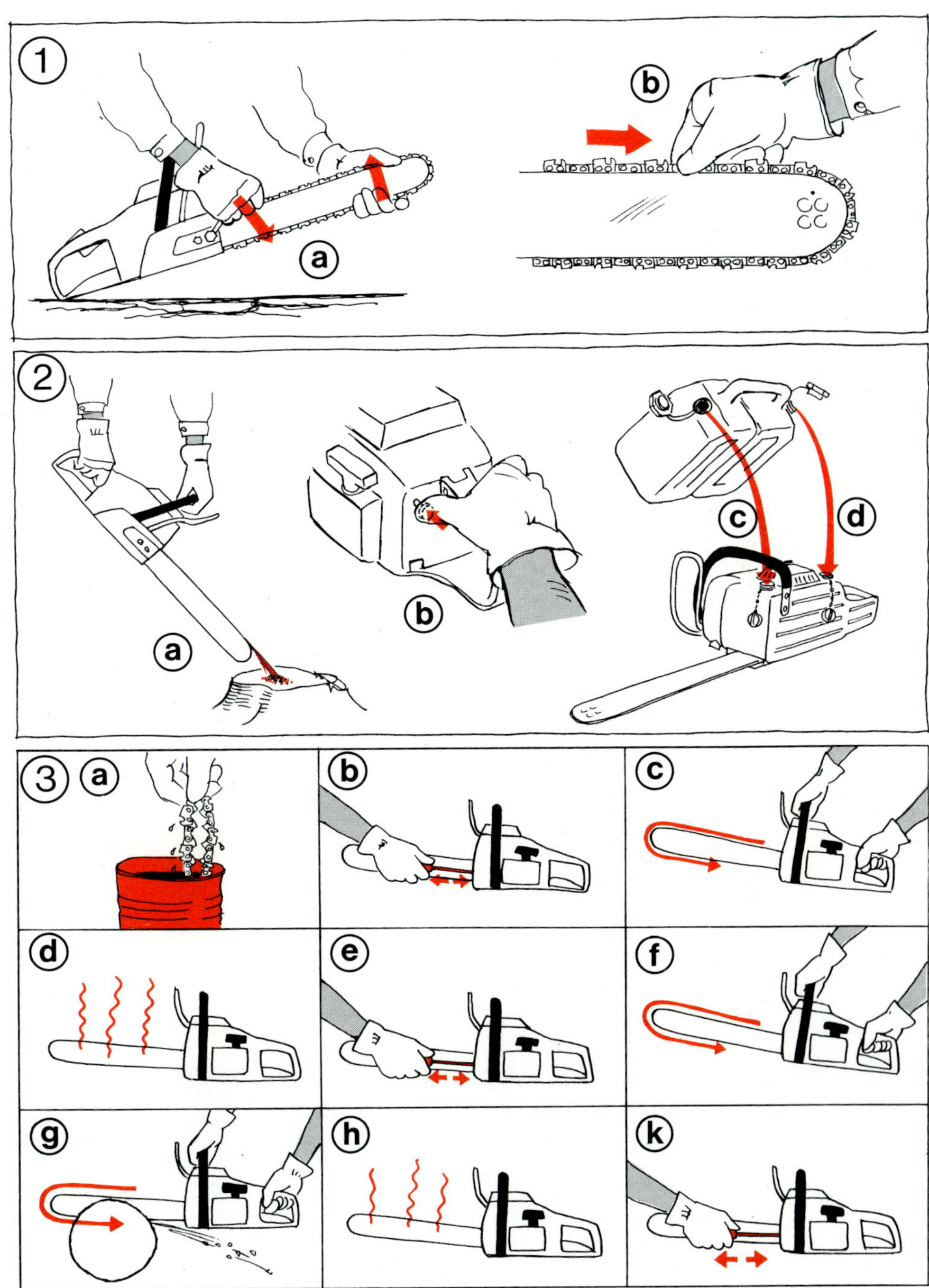
1
a
b
2
a
b
c
d
3
a
b
c
d
e
f
g
h
k

12.8 DISCARDING CHAINS, GUIDE BARS AND SPROCKETS

1. Saw chains should be discarded when:
   - the cutter has been filed back to the top plate length of about 4 mm (a);
   - the cutter link and the drive link are worn down to the rivets (b) because of wrong tension or insufficient oiling;
   - the chain remains stiff even if it is well oiled (c).

2. Sprockets or drive rims also wear out. Never use a worn sprocket or drive rim with a new chain, or a worn chain with a new sprocket or drive rim. Change the sprocket or drive rim when the tops of the teeth are worn out.

   Wear out two or three chains on the same sprocket by regularly changing the chains. Then continue with a new set of chains on a new sprocket or drive rim.

3. Replace a worn guide bar with a new one:
   - if the groove is worn out (a); or
   - if you find cracks or breaks (b).

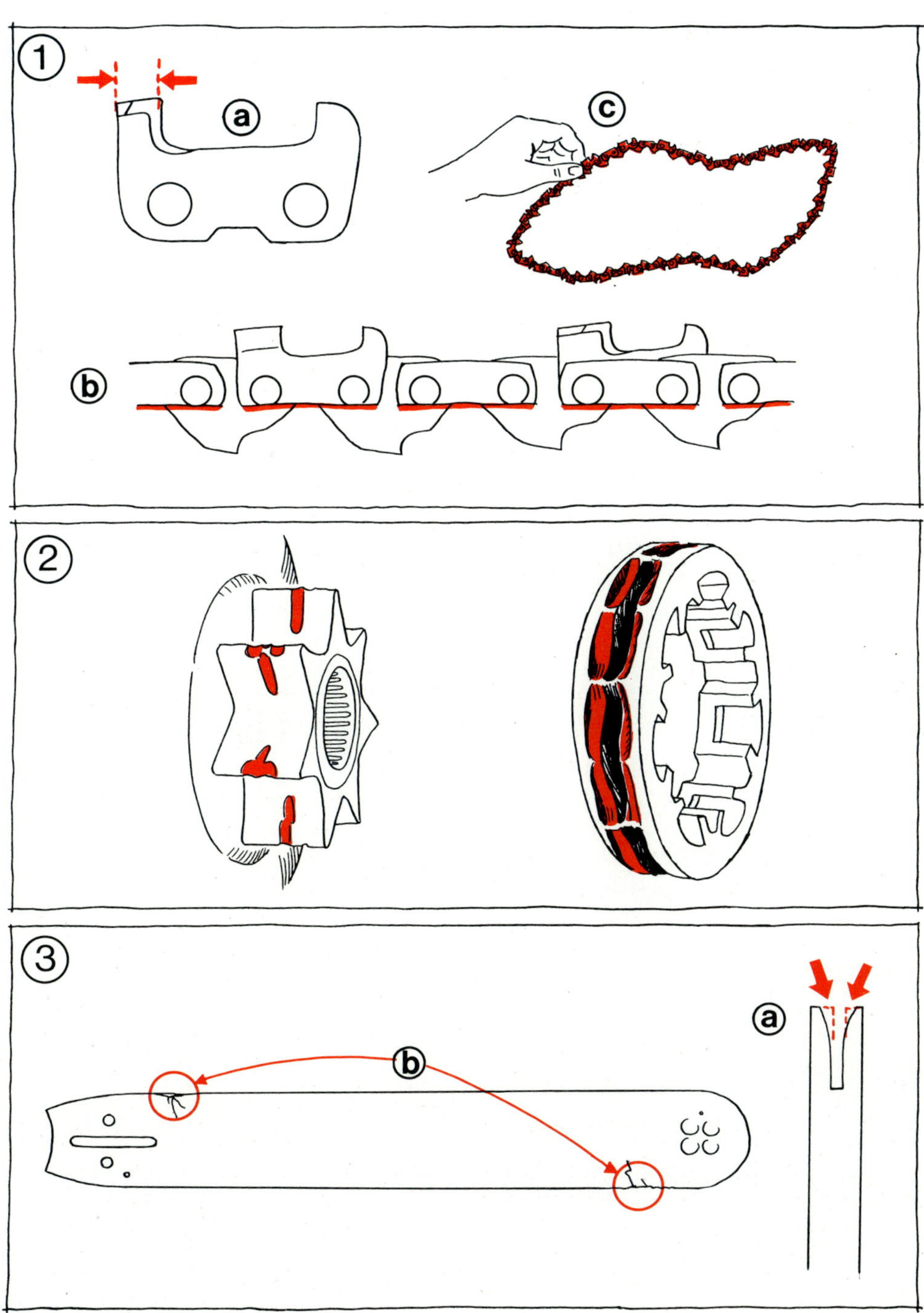
1
a
c
b
2
3
a
b

12.9 COMMON MISTAKES IN SAW CHAIN MAINTENANCE

The following mistakes are frequently made by inexperienced and untrained operators:

1. Cutter

Cutting edge slopes backwards (a). File was too thick or filing was done upwards. Cutter does not cut well.

Cutting edge shaped like a hook (b). File was too thin or filing was done downwards. Cutter does not cut smoothly, soon gets blunt, and the risk of kick-back is increased.

Side plate filing angle is too small (c). Cutter does not cut well, surface of cut is rough, chain may pinch.

Side plate filing angle too big (d). Cutter soon gets blunt, chain is worn out more rapidly.

Cutter length irregular (e). Chain does not cut smoothly. Chain runs to one side if all cutters on one side are longer than on the other side. The same happens if all cutters on one side have different filing angles than on the other side.

3. Depth gauge

If the depth gauge is too high (a) the chain does not cut well. If the depth gauge is too low (b), or if the front side is not rounded (c), the chain does not cut smoothly.

3. Tie strap and drive link

Excessive wear on those parts of the chain which are in contact with the guide bar and the sprocket are caused by wrong tension (too loose or too tight), insufficient chain oiling, cutting with a blunt or wrongly filed chain, or using a worn out sprocket or guide bar.

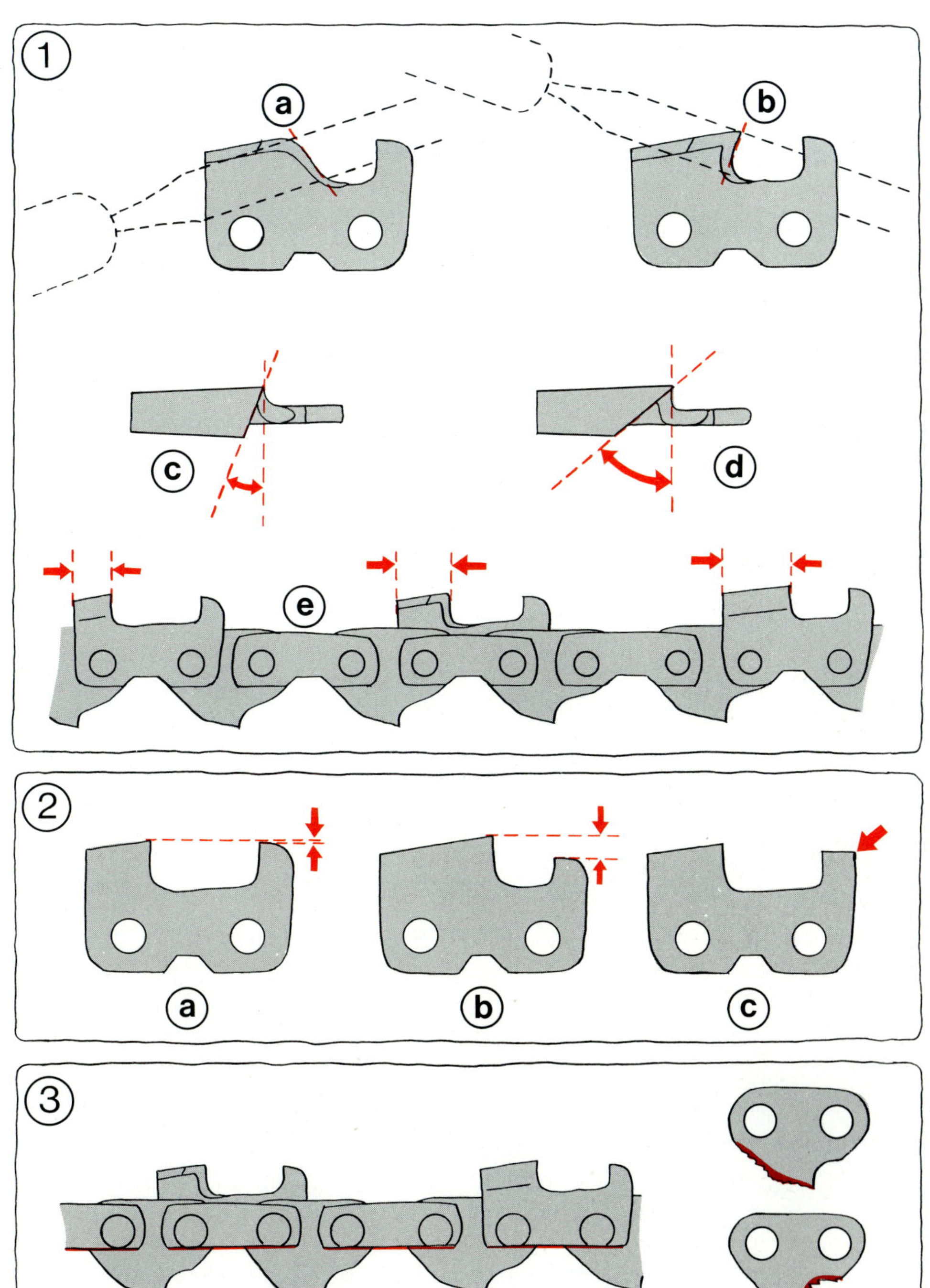
1
a
b
c
d
e
2
a
b
c
3

13. MAINTENANCE OF ENGINE

13.1 BASIC ENGINE MAINTENANCE

1. Air filter

A dirty air filter lowers the engine output and makes starting difficult. The air filter must be cleaned at least once a day. If there is a great deal of dust or dirt around, it must be cleaned more often.

Remove the filter cover.
Wash filter in soapy water or in petrol, using a small brush.

Always take one spare filter into the forest with you.

2. Cooling fan vanes and cylinder cooling fins

If the cooling fan (a) and the cylinder block and head fins (b) are covered with sawdust and dirt the engine will heat up and may break down. Therefore at least once a week the cooling fan vanes and the cooling fins must be cleaned with a stick or a screwdriver.

3. Spark plug

Clean the spark plug once a week. Check the electrode gap with a feeler gauge and adjust if necessary. The gap should be 0.5 mm.

4. Carburetor

Chainsaws are fitted with membrane carburetors. They have:

- a low-speed screw "L" (idle adjustment screw);
- a high-speed screw "H" (main adjustment screw); and
- a throttle adjusting screw "T" (idle speed regulating screw).

Carburetors are adjusted for maximum output with the lowest fuel consumption. If adjustment is necessary this should normally be done in the workshop. The experienced operator can do this in the forest. The engine must be warm and the air filter clean. First "L" and "H" are closed, then opened up by about one full turn. Next "T" is so adjusted that the engine neither stops nor the chain moves when the engine is idling. Make small adjustments on "L" until the engine speeds smoothly and quickly when the throttle is opened. Adjust "H" so that the engine runs smoothly during sawing.

1

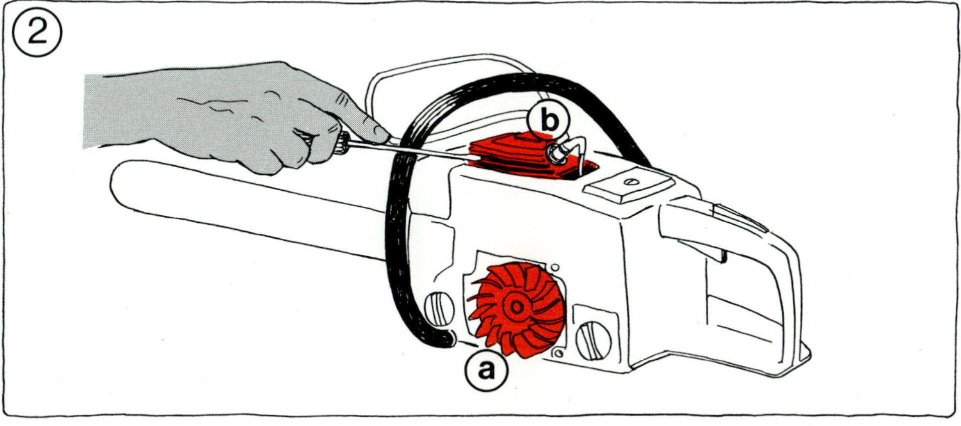
2
b
a

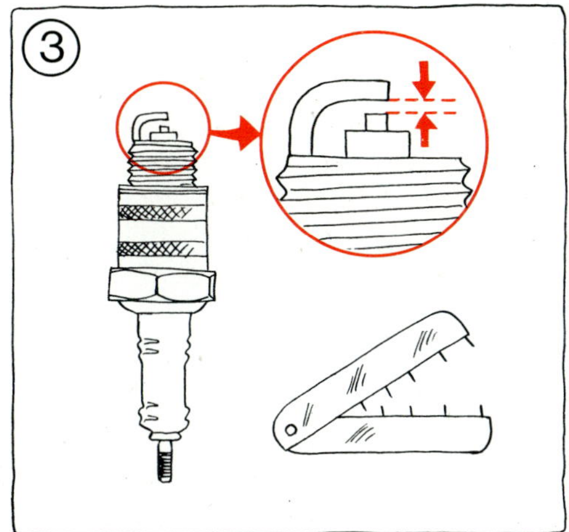
3

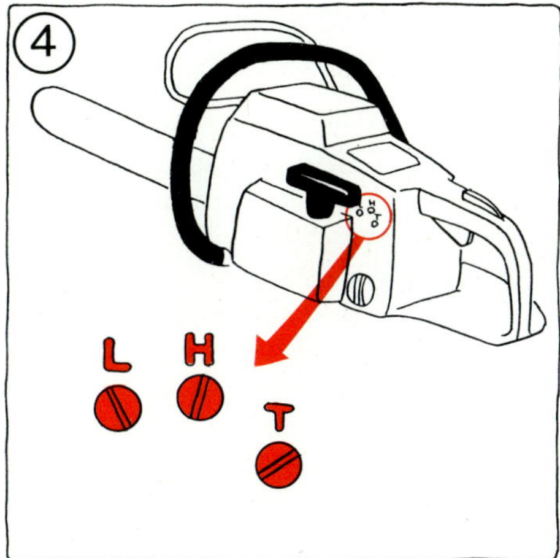
4
L
H
T

13.2 REPLACEMENT OF STARTER ROPE AND STARTER SPRING

Starter ropes or starter springs sometimes break. It is important that a spare starter rope and a spare starter spring are available at the work site. The operator should be trained to replace the broken parts himself.

1. To replace starter rope, first unscrew cover (a).
   Remove worn rope (b).
   Put new rope on rotor and anchor it safely (c).
   Pass the other end through cover and secure it to the handle with a double knot (d).
   Wind the rope on the rotor (e).
   Pull rope out about two turns (f).
   Hold rope rotor in position (g).
   Wind rope on rotor (h).
   The tension of the starter spring is correct when the rope rotor can still turn with the rope fully pulled out.
   Re-attach cover to saw (k).

2. To replace the starter spring, unscrew cover (a).
   Take the rope rotor off (b).
   Take the damaged spring out of spring housing (c).
   New springs are supplied in a cassette (d).
   Place the cassette over spring housing and push spring into proper position (e).
   If the spring has become loose, first put the outer loop into the spring housing and turn clockwise, ending with inner loop.
   Replace rope rotor (f).
   Adjust tension on starter spring and re-attach cover on the saw as described under 1 (e) to (k) above.

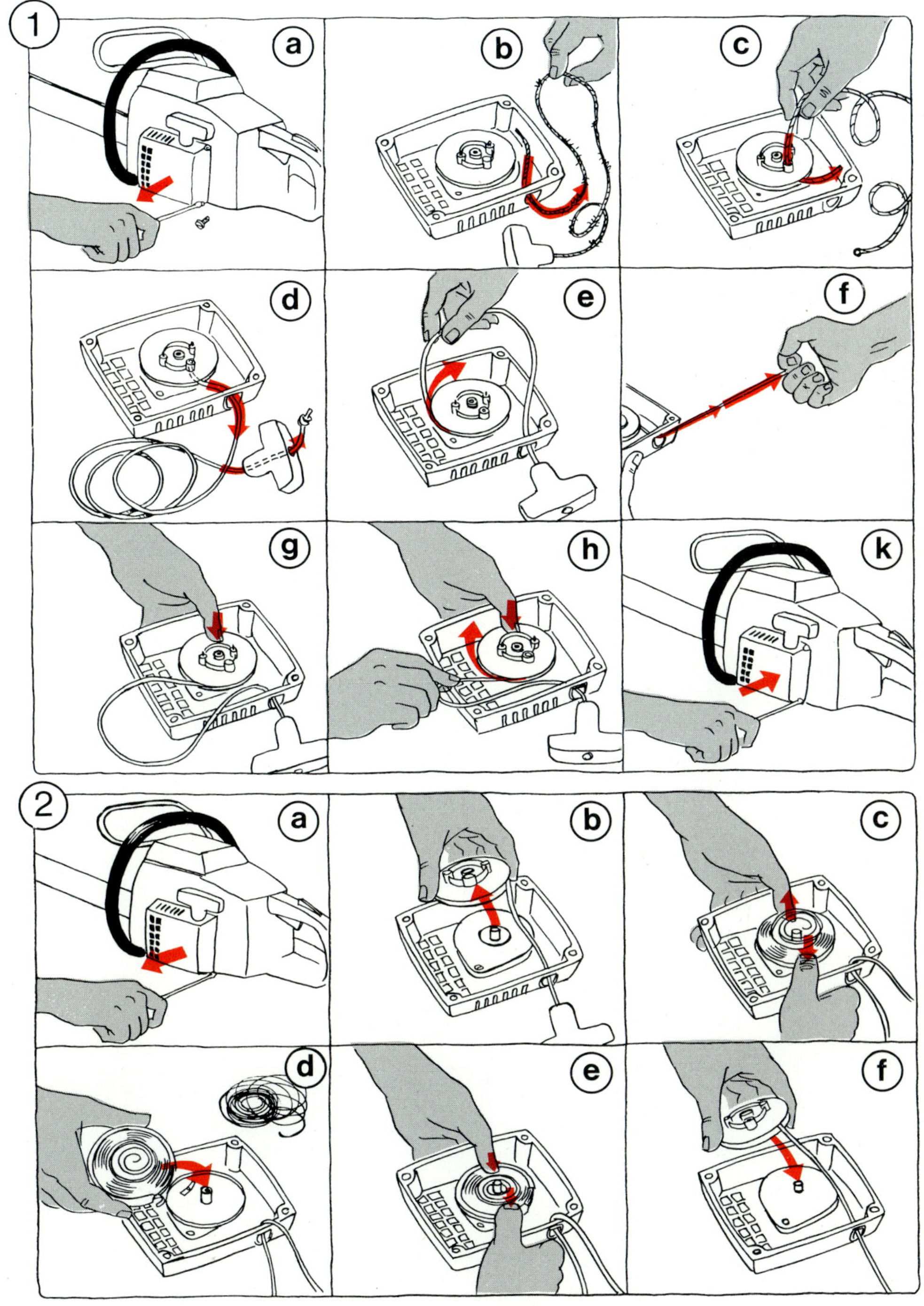
1
a
b
c
d
e
f
g
h
k
2
a
b
c
d
e
f

14. MAINTENANCE INTERVALS AND NECESSARY SPARE PARTS

14.1 DAILY MAINTENANCE OF CHAINSAW

To keep the chainsaw in good condition the following maintenance jobs must be done daily:

1. Chain

   Inspect and sharpen.

2. Guide bar

   Inspect, clean groove and oil holes, grease nose sprocket, check nuts. Turn the guide bar.

3. Engine part of the saw

   Clean carefully. Make sure that holes for air cooling intake are open.

4. Chain brake and front handle guard

   Inspect, clean and test.

5. Air filter

   Inspect and clean.

6. Screws, nuts, bolts

   Tighten.

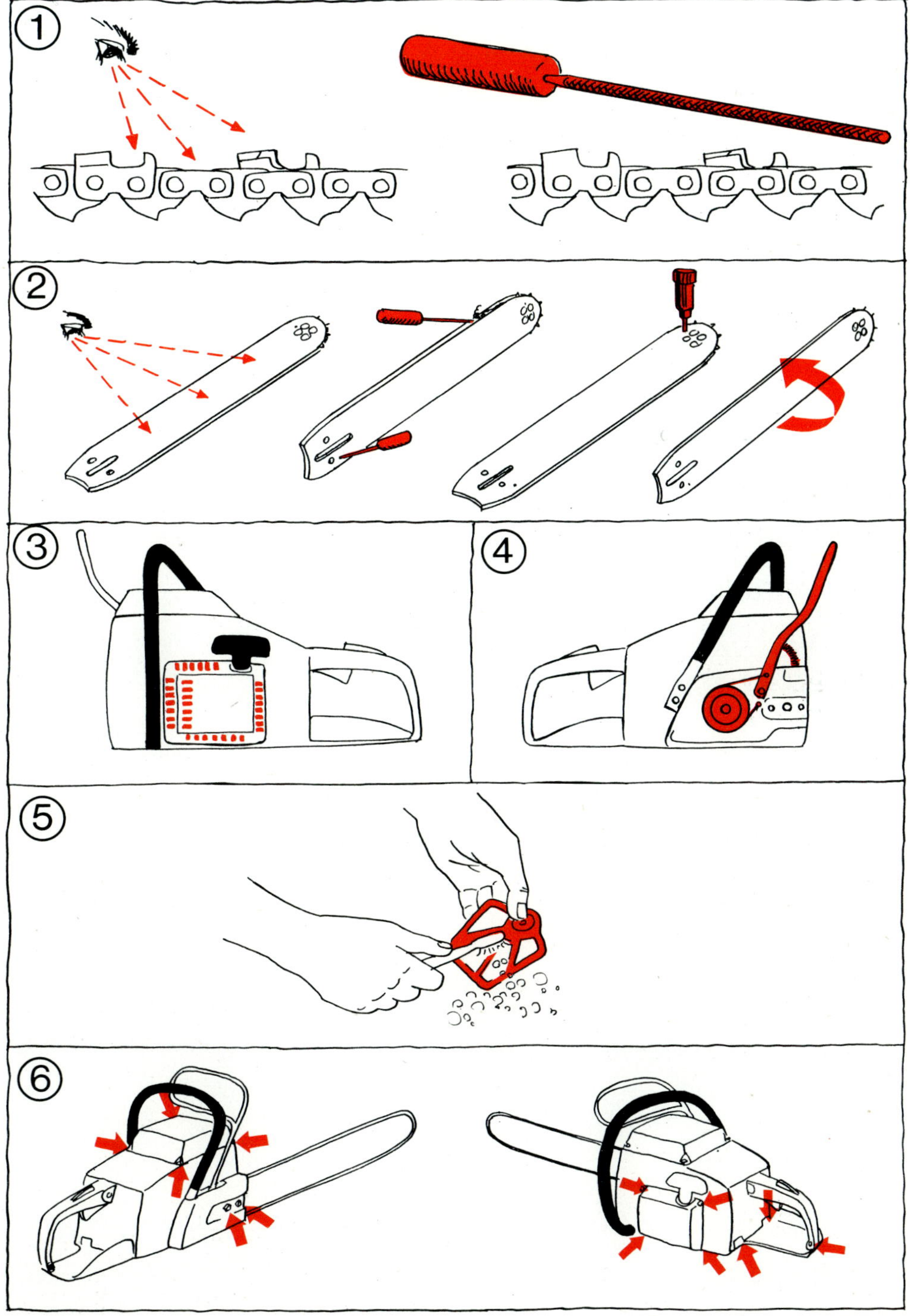
1
2
3
4
5
6

14.2 WEEKLY MAINTENANCE OF CHAINSAW

In addition to the daily maintenance jobs, the following jobs must be done at least once a week:

1. Chain

   Inspect and sharpen thoroughly.

2. Guide bar

   Remove burrs.

3. Sprocket (drive rim)

   Inspect, grease sprocket bearing.

4. Clutch

   Clean and inspect.

5. Cooling fan and cylinder fins

   Clean.

6. Spark plug

   Clean, inspect and adjust if necessary.

7. Starter

   Dismantle and clean, grease rope rotor bearing, replace rope if it is worn, adjust rope tension if necessary.

8. Oil and fuel filters

   Clean, check that oil reaches guide bar.

9. Muffler (exhaust)

   Clean.

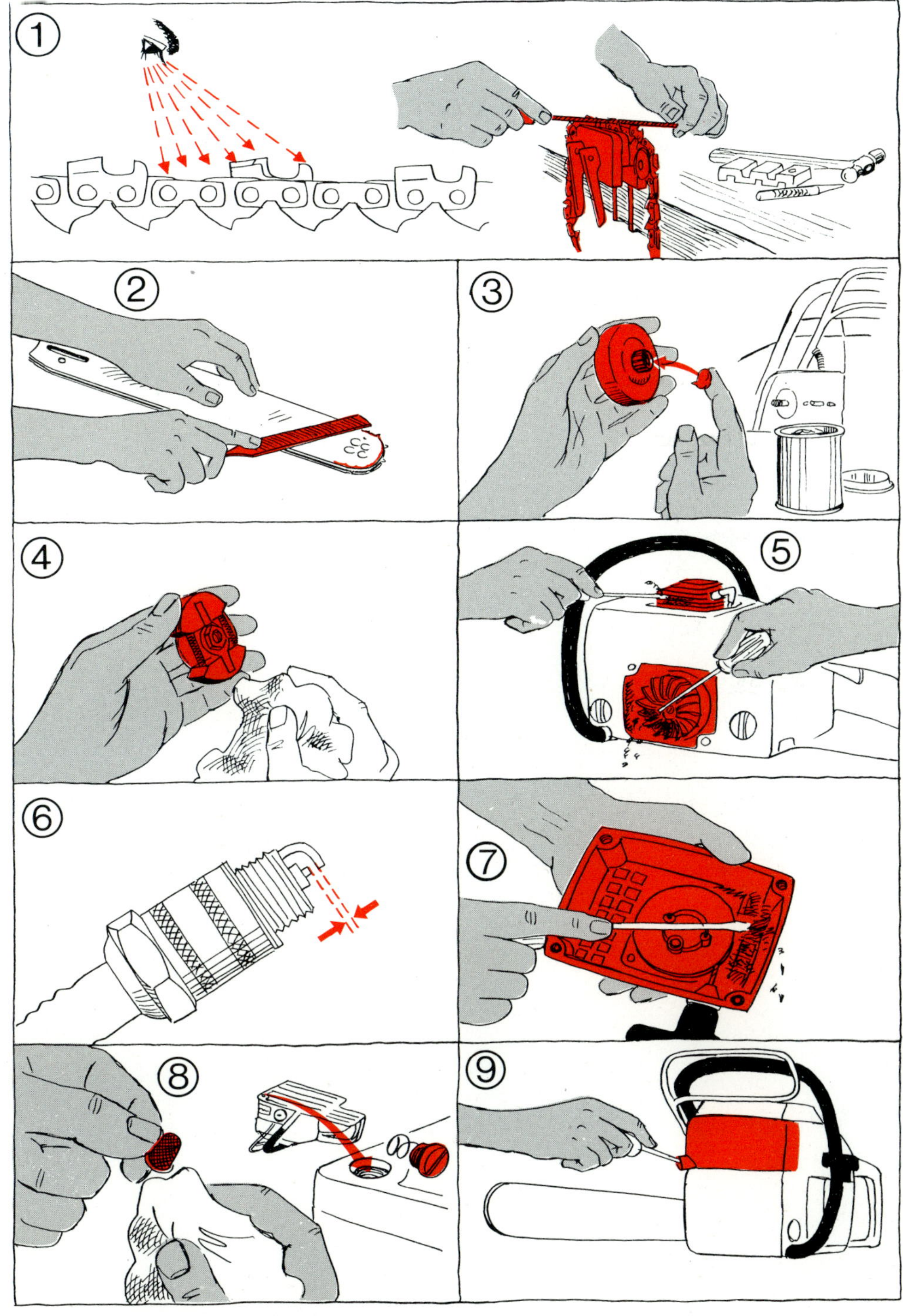
1
2
3
4
5
6
7
8
9

14.3 SPARE PARTS

The following spare parts should always be available at the working site:

1. Starter rope.

2. Starter spring.

3. Air filter.

4. Spark plug.

5. Guide bar nuts.

6. Casing and cover screws.

7. Chain.

8. Round files.

9. Flat files.

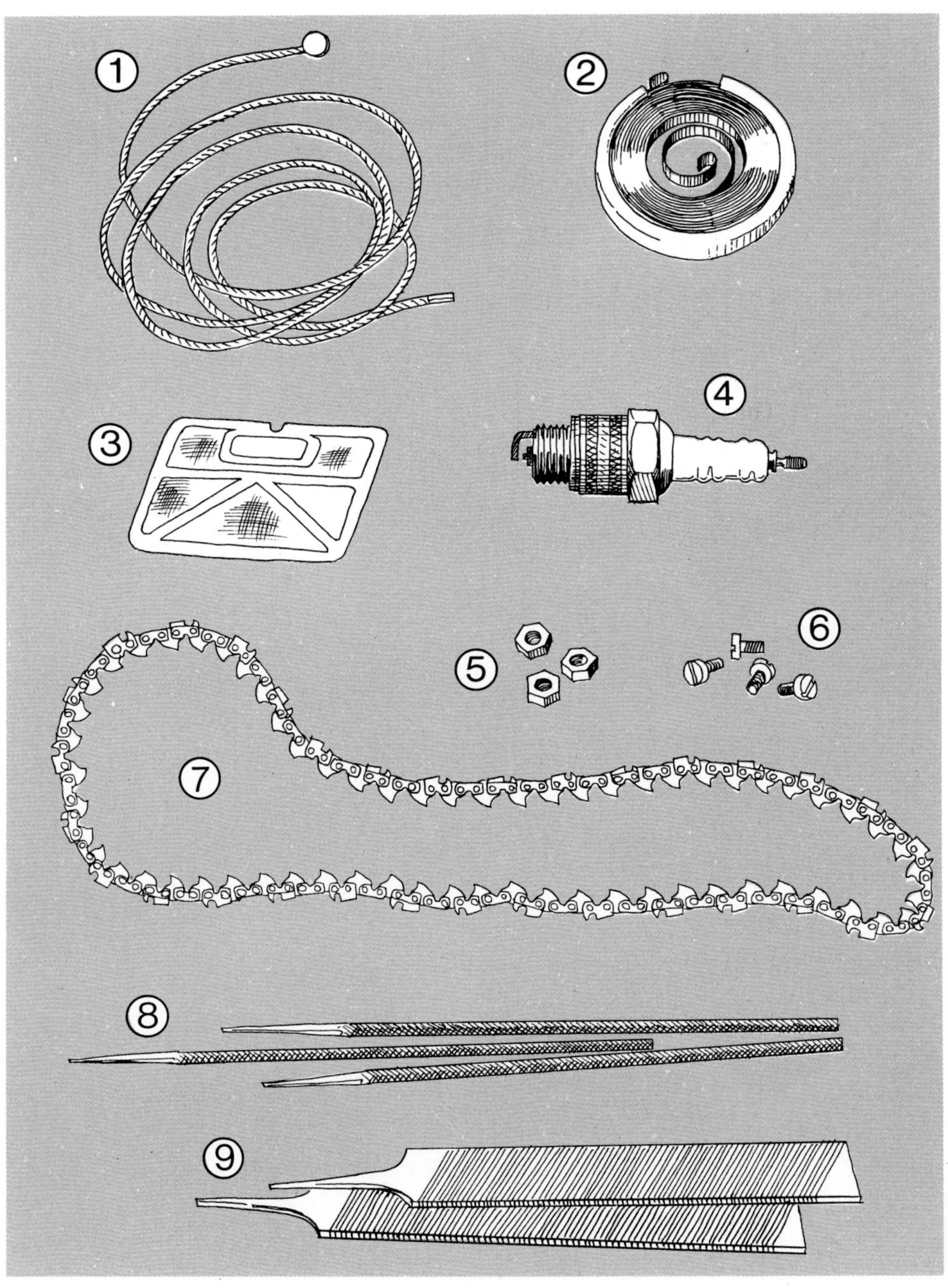
①
②
③
④
⑤
⑥
⑦
⑧
⑨

15. TRAINING OF CHAINSAW OPERATORS

Chainsaw operators need basic training and instruction to be able to work efficiently and safely without ruining the machine, the timber they are expected to produce, and eventually their health.

Basic training includes demonstrations and practical exercises in felling, delimbing and cross-cutting, in chain sharpening and in daily and weekly maintenance.

The length of basic training required will depend on the amount of experience the trainee already has. A minimum of two weeks would be needed for people with some experience.

Training requires qualified instructors who should only train small groups of workers at a time.

FAO and ILO will be glad to provide advice on the training of chainsaw operators and to assist in introducing training schemes.

FAO SALES AGENTS AND BOOKSELLERS

Algeria	Société nationale d'édition et de diffusion, 92, rue Didouche Mourad, Algiers.
Argentina	Editorial Hemisferio Sur S.A., Librería Agropecuaria, Pasteur 743, 1028 Buenos Aires.
Australia	Hunter Publications, 58A Gipps Street, Collingwood, Vic. 3066; Australian Government Publishing Service, Publishing Branch, P.O. Box 84, Canberra, A.C.T. 2600; and Australian Government Publications and Inquiry Centres in Canberra, Melbourne, Sydney, Perth, Adelaide and Hobart.
Austria	Gerold & Co., Buchhandlung und Verlag, Graben 31, 1011 Vienna.
Bangladesh	ADAB, 79 Road 11A, P.O. Box 5045, Dhanmondi, Dacca.
Belgium	Service des publications de la FAO, M.J. de Lannoy, 202, avenue du Roi, 1060 Brussels. CCP 000-0808993-13.
Bolivia	Los Amigos del Libro, Perú 3712, Casilla 450, Cochabamba; Mercado 1315, La Paz; René Moreno 26, Santa Cruz; Junín esq. 6 de Octubre, Oruro.
Brazil	Livraria Mestre Jou, Rua Guaipá 518, São Paulo 10; Rua Senador Dantas 19-S205/206, 20.031 Rio de Janeiro; PRODIL, Promoção e Dist. de Livros Ltda., Av. Venáncio Aires 196, Caixa Postal 4005, 90.000 Porto Alegre; A NOSSA LIVRARIA, CLS 104, Bloco C, Lojas 18/19, 70.000 Brasilia, D.F.
Brunei	SST Trading Sdn. Bhd., Bangunan Tekno No. 385, Jln 5/59, P.O. Box 227, Petaling Jaya, Selangor.
Canada	Renouf Publishing Co. Ltd, 2182 Catherine St. West, Montreal, Que. H3H 1M7.
Chile	Tecnolibro S.A., Merced 753, entrepiso 15, Santiago.
China	China National Publications Import Corporation, P.O. Box 88, Beijing.
Colombia	Litexsa Colombiana Ltda., Calle 55, Nº 16-44, Apartado Aéreo 51340, Bogotá D.E.
Costa Rica	Librería, Imprenta y Litografía Lehmann S.A., Apartado 10011, San José.
Cuba	Empresa de Comercio Exterior de Publicaciones, O'Reilly 407 Bajos entre Aguacate y Compostela, Havana.
Cyprus	MAM, P.O. Box 1722, Nicosia.
Czechoslovakia	ARTIA, Ve Smeckach 30, P.O. Box 790, 111 27 Praha 1.
Denmark	Munksgaard Boghandel, Norregade 6, 1165 Copenhagen K.
Dominican Rep.	Fundación Dominicana de Desarrollo, Casa de las Gárgolas, Mercedes 4, Apartado 857, Zona Postal 1, Santo Domingo.
Ecuador	Su Librería Cía. Ltda., García Moreno 1172 y Mejía, Apartado 2556, Quito; Chimborazo 416, Apartado 3565, Guayaquil.
El Salvador	Librería Cultural Salvadoreña S.A. de C.V., Calle Arce 423, Apartado Postal 2296, San Salvador.
Finland	Akateeminen Kirjakauppa, 1 Keskuskatu, P.O. Box 128, 00101 Helsinki 10.
France	Editions A. Pedone, 13, rue Soufflot, 75005 Paris.
Germany, F.R.	Alexander Horn Internationale Buchhandlung, Spiegelgasse 9, Postfach 3340, 6200 Wiesbaden.
Ghana	Fides Enterprises, P.O. Box 1628, Accra; Ghana Publishing Corporation, P.O. Box 3632, Accra.
Greece	G.C. Eleftheroudakis S.A., International Bookstore, 4 Nikis Street, Athens (T-126); John Mihalopoulos & Son, International Booksellers, 75 Hermou Street, P.O. Box 73, Thessaloniki.
Guatemala	Distribuciones Culturales y Técnicas " Artemis ", 5a. Avenida 12-11, Zona 1, Apartado Postal 2923, Guatemala.
Guinea-Bissau	Conselho Nacional da Cultura, Avenida da Unidade Africana, C.P. 294, Bissau.
Guyana	Guyana National Trading Corporation Ltd, 45-47 Water Street, P.O. Box 308, Georgetown.
Haiti	Librairie " A la Caravelle ", 26, rue Bonne Foi, B.P. 111, Port-au-Prince.
Hong Kong	Swindon Book Co., 13-15 Lock Road, Kowloon.
Hungary	Kultura, P.O. Box 149, 1389 Budapest 62.
Iceland	Snaebjörn Jónsson and Co. h.f., Hafnarstraeti 9, P.O. Box 1131, 101 Reykjavik.
India	Oxford Book and Stationery Co., Scindia House, New Delhi 110001; 17 Park Street, Calcutta 700016.
Indonesia	P.T. Sari Agung, 94 Kebon Sirih, P.O. Box 411, Djakarta.
Iran	Iran Book Co. Ltd, 127 Nadershah Avenue, P.O. Box 14-1532, Teheran.
Iraq	National House for Publishing, Distributing and Advertising, Jamhuria Street, Baghdad.
Ireland	The Controller, Stationery Office, Dublin 4.
Italy	Distribution and Sales Section, Food and Agriculture Organization of the United Nations, Via delle Terme di Caracalla, 00100 Rome; Libreria Scientifica Dott. Lucio de Biasio " Aeiou ", Via Meravigli 16, 20123 Milan; Libreria Commissionaria Sansoni S.p.A. " Licosa ", Via Lamarmora 45, C.P. 552, 50121 Florence.
Jamaica	Teacher Book Centre Ltd, 95 Church Street, Kingston.
Japan	Maruzen Company Ltd, P.O. Box 5050, Tokyo International 100-31.
Kenya	Text Book Centre Ltd, Kijabe Street, P.O. Box 47540, Nairobi.

FAO SALES AGENTS AND BOOKSELLERS

Korea, Rep. of	Eul-Yoo Publishing Co. Ltd, 112 Kwanchul-Dong, Chong-ro, P.O. Box Kwang-Whamoon No. 363, Seoul.
Kuwait	Saeed & Samir Bookstore Co. Ltd, P.O. Box 5445, Kuwait.
Luxembourg	Service des publications de la FAO, M.J. de Lannoy, 202, avenue du Roi, 1060 Brussels (Belgium).
Malaysia	SST Trading Sdn. Bhd., Bangunan Tekno No. 385, Jln 5/59, P.O. Box 227, Petaling Jaya, Selangor.
Mauritius	Nalanda Company Limited, 30 Bourbon Street, Port Louis.
Mexico	Dilitsa S.A., Puebla 182-D, Apartado 24-448, Mexico 7, D.F.
Morocco	Librairie " Aux Belles Images ", 281, avenue Mohammed V, Rabat.
Netherlands	Keesing Boeken B.V., Hondecoeterstraat 16, 1017 LS Amsterdam.
New Zealand	Government Printing Office: Government Bookshops at Rutland Street, P.O. Box 5344, Auckland; Alma Street, P.O. Box 857, Hamilton; Mulgrave Street, Private Bag, Wellington; 130 Oxford Terrace, P.O. Box 1721, Christchurch; Princes Street, P.O. Box 1104, Dunedin.
Nigeria	University Bookshop (Nigeria) Limited, University of Ibadan, Ibadan.
Norway	Johan Grundt Tanum Bokhandel, Karl Johansgate 41-43, P.O. Box 1177 Sentrum, Oslo 1.
Pakistan	Mirza Book Agency, 65 Shahrah-e-Quaid-e-Azam, P.O. Box 729, Lahore 3.
Panama	Distribuidora Lewis S.A., Edificio Dorasol, Calle 25 y Avenida Balboa, Apartado 1634, Panama 1.
Paraguay	Agencia de Librerías Nizza S.A., Tacuarí 144, Asunción.
Peru	Librería Distribuidora " Santa Rosa ", Jirón Apurímac 375, Casilla 4937, Lima 1.
Philippines	The Modern Book Company Inc., 926 Rizal Avenue, P.O. Box 632, Manila.
Poland	Ars Polona, Krakowskie Przedmiescie 7, 00-068 Warsaw.
Portugal	Livraria Bertrand, S.A.R.L., Rua João de Deus, Venda Nova, Apartado 37, Amadora; Livraria Portugal, Dias y Andrade Ltda., Rua do Carmo 70-74, Apartado 2681, 1117 Lisbon Codex; Edições ITAU, Avda. da República 46/A-r/c Esqdo., Lisbon 1.
Romania	Ilexim, Calea Grivitei N° 64-66, B.P. 2001, Bucharest.
Saudi Arabia	University Bookshop, Airport Street, P.O. Box 394, Riyadh.
Senegal	Librairie Africa, 58, avenue Georges Pompidou, B.P. 1240, Dakar.
Singapore	MPH Distributors (S) Pte. Ltd, 71/77 Stamford Road, Singapore 6; Select Books Pte. Ltd, 215 Tanglin Shopping Centre, Tanglin Road, Singapore 1024; SST Trading Sdn. Bhd., Bangunan Tekno No. 385, Jln 5/59, P.O. Box 227, Petaling Jaya, Selangor.
Somalia	" Samater's ", P.O. Box 936, Mogadishu.
Spain	Mundi Prensa Libros S.A., Castelló 37, Madrid 1; Librería Agrícola, Fernando VI 2, Madrid 4.
Sri Lanka	M.D. Gunasena & Co. Ltd, 217 Olcott Mawatha, P.O. Box 246, Colombo 11.
Sudan	University Bookshop, University of Khartoum, P.O. Box 321, Khartoum.
Suriname	VACO n.v. in Suriname, Dominee Straat 26, P.O. Box 1841, Paramaribo.
Sweden	C.E. Fritzes Kungl. Hovbokhandel, Regeringsgatan 12, P.O. Box 16356, 103 27 Stockholm.
Switzerland	Librairie Payot S.A., Lausanne et Genève; Buchhandlung und Antiquariat Heinimann & Co., Kirchgasse 17, 8001 Zurich.
Tanzania	Dar es-Salaam Bookshop, P.O. Box 9030, Dar es-Salaam; Bookshop, University of Dar es-Salaam, P.O. Box 893, Morogoro.
Thailand	Suksapan Panit, Mansion 9, Rajadamnern Avenue, Bangkok.
Togo	Librairie du Bon Pasteur, B.P. 1164, Lomé.
Trinidad and Tobago	The Book Shop, 22 Queens Park West, Port of Spain.
Tunisia	Société tunisienne de diffusion, 5, avenue de Carthage, Tunis.
United Kingdom	Her Majesty's Stationery Office, 49 High Holborn, London WC1V 6HB (callers only); P.O. Box 569, London SE1 9NH (trade and London area mail orders); 13a Castle Street, Edinburgh EH2 3AR; 41 The Hayes, Cardiff CF1 1JW; 80 Chichester Street, Belfast BT1 4JY; Brazennose Street, Manchester M60 8AS; 258 Broad Street, Birmingham B1 2HE; Southey House, Wine Street, Bristol BS1 2BQ.
United States of America	UNIPUB, 345 Park Avenue South, New York, N.Y. 10010.
Uruguay	Librería Editorial Juan Angel Peri, Alzaibar 1328, Casilla de Correos 1755, Montevideo.
Venezuela	Blume Distribuidora S.A., Gran Avenida de Sabana Grande, Residencias Caroni, Local 5, Apartado 70.017, Caracas.
Yugoslavia	Jugoslovenska Knjiga, Trg. Republike 5/8, P.O. Box 36, 11001 Belgrade; Cankarjeva Zalozba, P.O. Box 201-IV, 61001 Ljubljana; Prosveta, Terazije 16, P.O. Box 555, 11001 Belgrade.
Zambia	Kingstons (Zambia) Ltd, Kingstons Building, President Avenue, P.O. Box 139, Ndola.
Other countries	Requests from countries where sales agents have not yet been appointed may be sent to: Distribution and Sales Section, Food and Agriculture Organization of the United Nations, Via delle Terme di Caracalla, 00100 Rome, Italy.